SATURN'S MOON TITAN

From 4.5 billion years ago to the present

COVER IMAGE: A near-infrared (938nm) image of Titan, taken by Cassini's camera from 500,000km distance during the last few months of the mission. A bright streak of methane cloud catches the eye, while irregular areas towards the bottom are giant dunefields of dark sand. The darkest splotches near the top are Titan's seas of liquid methane.
(NASA/JPL/SSI)

© Ralph Lorenz 2020

All rights reserved. No part of this publication may be reproduced or stored in a retrieval system or transmitted, in any form or by any means, electronic, mechanical, photocopying, recording or otherwise, without prior permission in writing from Haynes Publishing.

First published in June 2020

A catalogue record for this book is available from the British Library.

ISBN 978 1 78521 643 5

Library of Congress control no. 2020931442

Published by Haynes Publishing,
Sparkford, Yeovil, Somerset BA22 7JJ, UK.
Tel: 01963 440635
Int. tel: +44 1963 440635
Website: www.haynes.com

Haynes North America Inc.,
859 Lawrence Drive, Newbury Park,
California 91320, USA.

Printed in China.

SATURN'S MOON TITAN

From 4.5 billion years ago to the present

Owners' Workshop Manual

An insight into the workings and exploration of the most Earth-like world in the outer solar system

Ralph Lorenz

Contents

6	Preface	
8	Titan: The moon that would be a planet	
20	Titan before Cassini-Huygens	
	Voyager encounters	25
	First glimpses of the surface	28
34	Titan unveiled	
	Huygens descent	40
	Methane and hydrogen profiles	46
	Haze	47
	Uncovering Titan	48
60	Evolution and interior	
	Mountains and tectonics	68
	Craters	70
	Balance of power	76
	Xanadu	77
	Titan's formation and orbit	79
80	Rivers and dunes	
	Rivers	82
	Canyonlands	86
	Dunes	91
98	Lakes and seas	
	Tides	111
	Waves	114
118	Restless atmosphere	
	Greenhouse structure	120
	Cloud formation and rainfall	122
	Temperature variations and winds	128
	Twilight and visibility	130
	Stratospheric circulation	131
136	Chemistry and space interaction	
	Photochemistry	144
	Oxygen chemistry	150
	Dance of gravity	151
152	Future exploration	
	After Cassini and before Dragonfly	166
168	Life on Titan?	
	Alien life chemistry	173
	Humans on Titan	174
	The distant future	178
180	Maps	
184	List of place names	
193	Further reading	
194	Index	

OPPOSITE A quasi-true-colour image of Titan. The bright and dark features on the surface are actually sensed in the near-infrared at 938nm. The faint patches at the lower-left are dunefields and the dark blotches at the top are Titan's north polar lakes and seas. *(SSI/NASA/JPL)*

Preface

By United Nations Treaty a planetary body cannot be owned by anyone, so the notion of an 'Owners' Workshop Manual' for a world struck me as a little odd. Nevertheless, I was excited to have the opportunity to work with Haynes on this book. The affordable, all-colour wide format is an excellent vehicle to lay out the workings of the Saturnian moon, Titan, and a couple of years after the end of the Cassini mission is the perfect time to sum up what we know.

I have striven to make this volume complement, rather than duplicate, my *Cassini-Huygens Owners' Workshop Manual*, and the reader is referred to that for the intimate details of those spacecraft and their operation.

This book is not a textbook; it is both too broad and insufficiently rigorous for that. But I have not held back on the use of scientific jargon because, in this day and age, the Internet makes it easy to learn more about unfamiliar concepts. My goal has been to broaden horizons and share the many dimensions that make Titan interesting. An astrobiologist may not know how dune morphology relates to winds, while someone who delights in geology may not have heard of airglow. I hope this book will at least help the reader to appreciate, through comparisons with features on Earth or elsewhere, all these fascinating aspects of Titan.

As the book neared completion in mid-2019, NASA announced that its next billion-dollar class New Frontiers space mission would be 'Dragonfly', a rotorcraft relocatable lander for Titan. I was delighted because I played a central role in devising that proposal. Dragonfly is planned to launch in 2026, and to arrive at Titan in 2034, exactly one Titan year after the Cassini mission delivered the Huygens probe in 2005. I started my career on the early design of the Huygens probe, and I hope to see Dragonfly accomplish several years of operations before I retire. My entire professional life has been devoted to the exploration of Titan, and I hope this book will help the reader understand why this magnificent world deserves such attention.

I am grateful to Steve Rendle at Haynes for commissioning me to write this volume, and also for the deft editing by Titan veteran David M. Harland. I also thank my many colleagues who provided graphical material and have shared the Titan adventure with me, not least Dr. Elizabeth Turtle.

Shannon MacKenzie provided enthusiastic and helpful feedback on the manuscript as well as several images.

Ralph Lorenz
Johns Hopkins Applied Physics Laboratory (APL)
Laurel, MD, USA.

OPPOSITE A near-infrared view of Titan by Cassini centred on Xanadu, taken in late 2004 just after Cassini's arrival in southern summer. There are bright clouds at the south pole and the Shangri-La dunefields are to the west. *(SSI/JPL/NASA)*

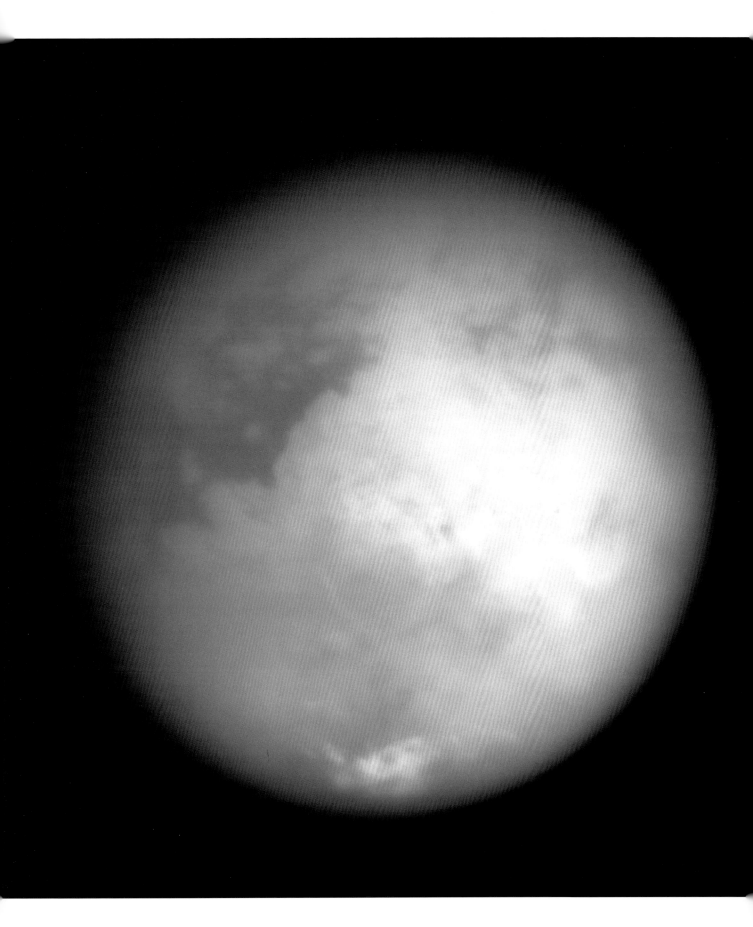

Chapter One

Titan: The moon that would be a planet

Titan, bigger than the planet Mercury, is the only moon in the solar system to possess a dense atmosphere. Its climate and landscape bear comparison with the terrestrial planets (Earth, Mars and Venus) as well as icy bodies such as the large Jovian moon Ganymede and the outer dwarf planet Pluto. Titan has even served as a prototype 'exoplanet'.

OPPOSITE Titan by Cassini's camera in approximately true colour, darker and redder than its parent planet, which in turn is darker and redder than the icy rings. On the right are the filigree shadows of the ring cast on Saturn. *(NASA/JPL/SSI)*

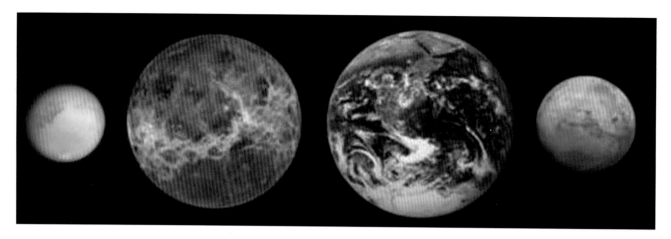

ABOVE Titan and the terrestrial planets: the worlds that have climates. Titan, like Venus, rotates relatively slowly and has an opaque atmosphere. Just like Earth and Mars, the tilt of its equator gives Titan seasons. But like only the Earth, Titan has a hydrological cycle with clouds and rain onto the surface. *(Author)*

BELOW Titan only just missed out to Ganymede for being the largest moon in the solar system. Also, whereas Jupiter has four similar-sized satellites, Titan completely dominates the Saturnian system, being ten times more massive than its siblings, which themselves are roughly the same size as the moons of Uranus. *(NASA)*

All worlds are unique, but Titan is more unique than most. It happens to orbit Saturn, and for that reason it cannot be considered a planet in its own right. However, as the only moon in the solar system to possess a thick atmosphere, one that happens to be composed predominantly of molecular nitrogen, as does ours, Titan has more of the trappings of a planet than many actual planets!

Titan's atmosphere is meagre compared

with the vast hydrogen-rich envelopes of the giant planets Jupiter and Saturn, and the ice giants Uranus and Neptune, but these planets do not have accessible solid surfaces, no place on which one can stand and have a defining vista; *no landscape*.

In so far as Titan's atmosphere is dense, optically thick and slowly rotating, it has a degree of commonality with that of Venus, but because that planet's equator is aligned with its orbital plane it does not have seasons. In contrast, Titan's equator is inclined by about 25°, much like Earth and Mars, causing profound seasons at its high latitudes. From Titan's north polar seas, the Sun lingers low in the summer sky, but does not set, and methane rainclouds puff up from the heated surface. Yet each pole dwells for several Earth years in long winter darkness, with exotic chemicals condensing in multiple layers of cloud crystals high in the atmosphere above.

At present, these seas and rain are features shared only by our own planet. They may have been present on Mars, and perhaps Venus, in earlier epochs, but only Earth supports standing bodies of liquid and a hydrological cycle in which the liquid evaporates to form clouds, and then rains back down to form river channels which tumble cobbles and pebbles into rounded shapes. The fact that on Titan the liquid is methane and on Earth it is water is a mere detail – the physical processes, and it seems, the resultant landforms, are the same.

Titan's climate, mirroring that of Earth but with different temperatures and compositions, also serves as a prototype for those of bodies now being found around other stars – exoplanets and exomoons. Indeed, many of the astronomical observations used to study these worlds, and the mathematical models used to understand them, were first tried out on Titan.

The methane in Titan's atmosphere provides the starting point for a complex organic chemistry driven by sunlight which

CONTINUED ON PAGE 16 ▶

BELOW Titan (here in false colour) to scale in the Saturnian system. Its orbit around Saturn is 20 times the diameter of the planet itself. (D. Seal/Caltech/JPL)

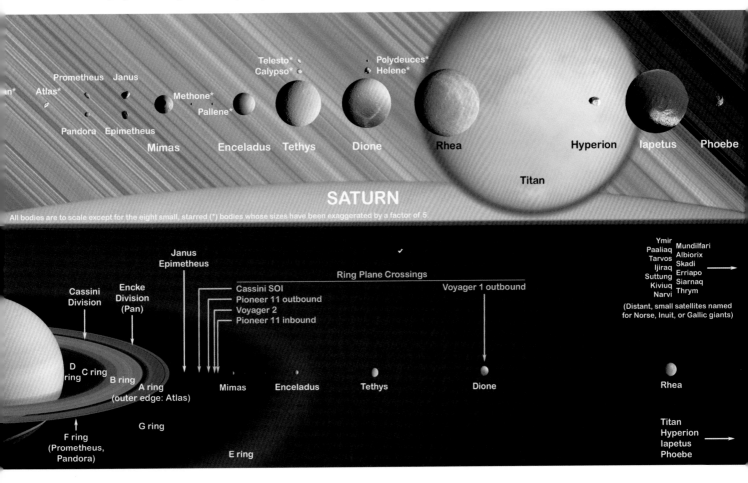

11

TITAN: THE MOON THAT WOULD BE A PLANET

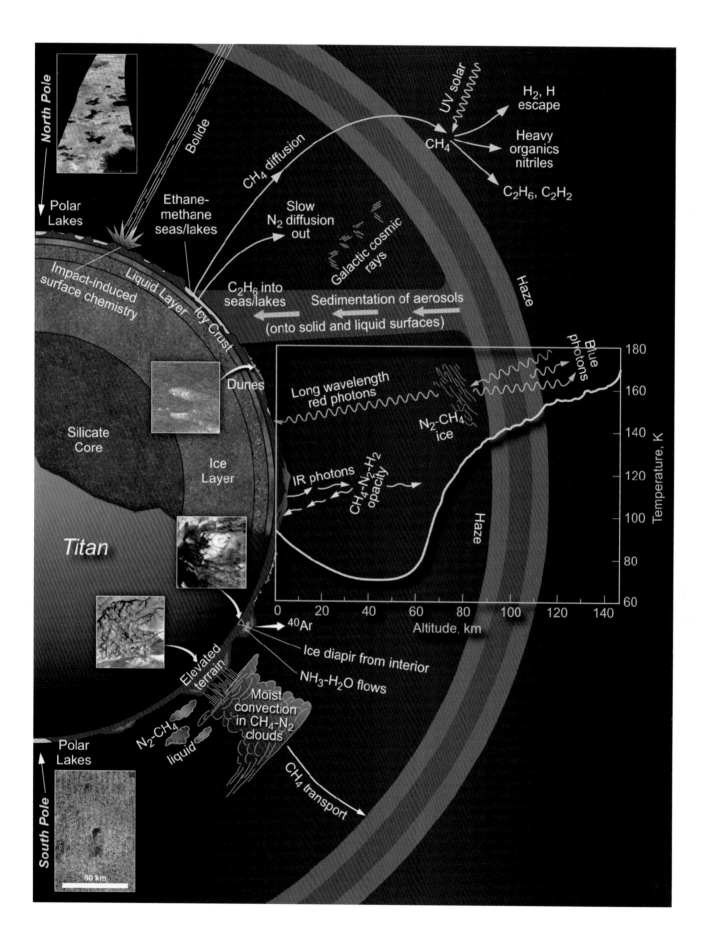

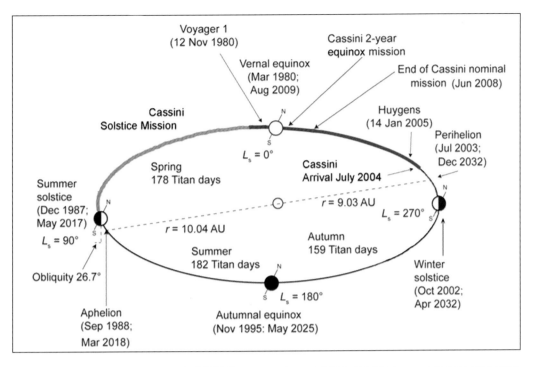

LEFT The seasons in Titan's 29.5-year annual cycle. Cassini arrived shortly after southern midsummer. The eccentricity of Saturn's orbit around the Sun means that Titan's seasons are uneven, making southern summer presently shorter and hotter than the north. *(Author)*

OPPOSITE Titan as an active world. The meteorological/hydrological cycle that we observe in the atmosphere is part of a wider interaction between Titan and its space environment, and between the surface and interior of this most diverse world. *(NASA/JPL)*

LEFT Titan for photographers. Taken by Cassini's camera at 940nm (the near-infrared wavelength widely used for TV remote controls) where the haze is thinner than in visible light, it features the dark equatorial dunefields. A faint dark circle can be seen at the edge of Titan's disc – a detached haze layer. The diagonal line across the middle of the image is Saturn's ring, and at the bottom-right are the shadows of the ring on Saturn. *(SSI/NASA/JPL)*

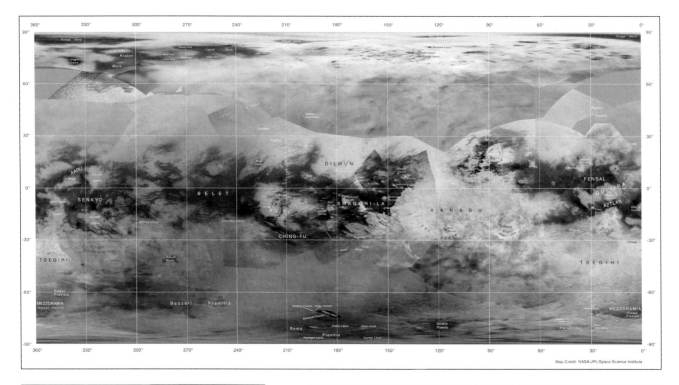

Map Credit: NASA/JPL/Space Science Institute

Titan | Moon at Similar Scale

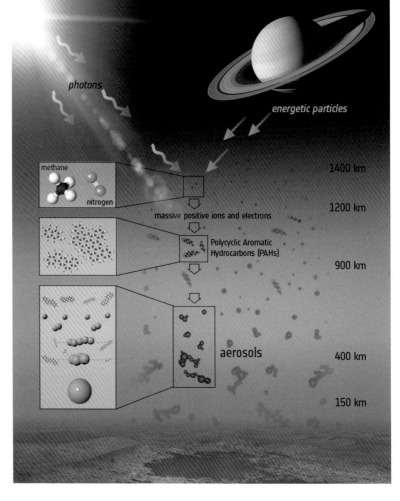

OPPOSITE TOP Titan for cartographers. This map of Titan's surface in the near-infrared (940nm) was compiled in 2016 from thousands of Cassini images. Brightness discontinuities indicate seams where observations taken under different illumination conditions have not been fully corrected. Names for many of the features have been approved by the International Astronomical Union. *(SSI/NASA/JPL)*

RIGHT Titan for romantics. A 5μm image of Titan with the Sun behind it. The bright spot is a lake (Jingpo Lacus) near the north pole. Because the surface of the lake is as flat as a millpond it shows a mirror-like ('specular') reflection of the Sun. *(NASA/JPL/U. Arizona/DLR)*

RIGHT Titan for geologists. A showcase of radar images of (a) lakes, (b) impact crater Menrva and river channels Elivagar Flumina, (c) the dunefields in Belet, (d) river channels in Xanadu, and (e) impact crater Sinlap. *(Author)*

OPPOSITE BOTTOM LEFT Titan for explorers. The view on the left was taken by the Huygens probe from about knee-height in 2005. It was a greyscale image, but has been authentically coloured based on data from the probe's spectrometers. The right image is from the lunar surface at a similar scale. The white oval at the bottom-right of the Huygens image is saturated due to strong lamp illumination. *(E. Karkoschka, U. Arizona/ESA/NASA)*

OPPOSITE BOTTOM RIGHT Titan for chemists. A myriad of complex carbon, nitrogen and hydrogen compounds are formed in Titan's atmosphere, eventually being deposited on the surface. *(ESA/ATG Medialab)*

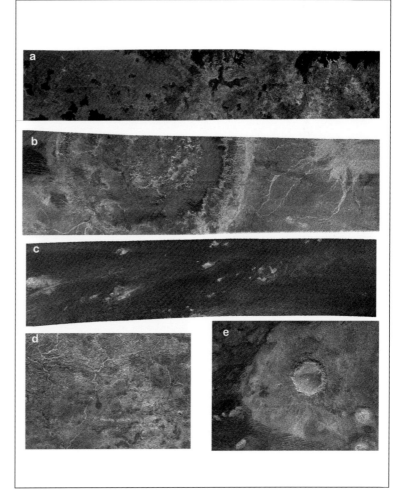

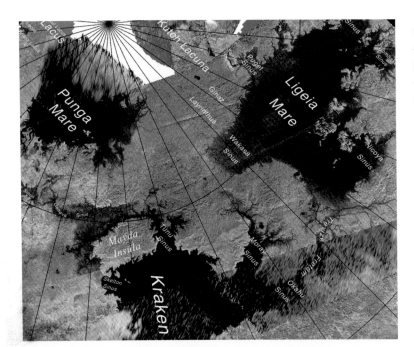

LEFT Titan for sailors and oceanographers. At ~400km across, Ligeia Mare is Titan's second-largest methane sea. Kraken Mare sprawls over ~1,000km and projects tidal currents through narrow straits. *(USGS)*

may mirror aspects of that on the early Earth. How important Titan-like photochemistry may have been in the origins of life on Earth is a matter of speculation, the evidence having been obliterated (or eaten!) in the tumultuous aeons of our restless, living planet. But on Titan these processes remain in deep freeze, ready to be explored.

And Titan presents an outstanding opportunity for exploration. Although Titan's great distance from Earth is indeed inconvenient – it takes spacecraft several years to get there, and even light and radio signals take more than an hour – Titan's thick atmosphere serves as a cushion to slow spacecraft from their rocketing interplanetary speeds. Thus spacecraft can use heat shields and parachutes to enter into the atmosphere and land safely without the dangerous complication of rocket engines. Titan is therefore a much easier surface to explore than that of the Jovian moon Europa, for example.

All of this gives Titan a broad scientific appeal beyond any individual discipline. Its appeal is perhaps second only to Mars. And it is within our reach. Unlike torrid Venus, where complex spacecraft can survive only an hour or two, a lander for Titan's frigid atmosphere will require only a nuclear heat source. As a result, NASA has named Titan as the destination for one of its forthcoming endeavours in planetary exploration, named Dragonfly.

Titan has even entered the popular imagination, with dozens of science fiction novels now set there. In some (e.g. *The Puppet*

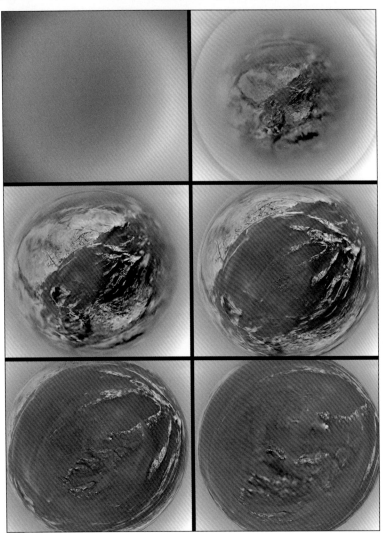

LEFT Titan for parachutists. Images from the Huygens probe are mosaicked together and coloured with data from its spectrometers, and the resultant scene projected in a fish-eye manner similar to a GoPro™ camera from ever-lower altitudes during the parachute descent. *(NASA/ESA/U.Arizona)*

> There were regions of Titan where a man needed little more than an oxygen mask and a simple thermofoil suit to move around in the open. To everyone's great surprise, Titan had turned out to be the most hospitable place in the solar system, next to Earth itself.
>
> **Arthur C. Clarke** *Imperial Earth*, 1975

Masters) Titan is just mentioned in passing, being an unknown moon as mysterious as any other, but in many others, and especially in more recent stories, its unique and exotic environment is exploited for clever plot devices. Remarkably, imagination has sometimes anticipated reality: the dunefields in the Japanese anime *Cowboy Bepop: Attack on Titan*, for example, or the three seas in *The Sirens of Titan*, and the methane monsoon in *Imperial Earth*. Perhaps there is no greater testament to Titan's recognition by popular culture than the iconic view in the 2009 film *Star Trek* of the *USS Enterprise* emerging from Titan's haze, with Saturn in the background.

The aim of this book is to lay out what we now know about Titan, and how its diverse aspects interact. In short, how Titan works.

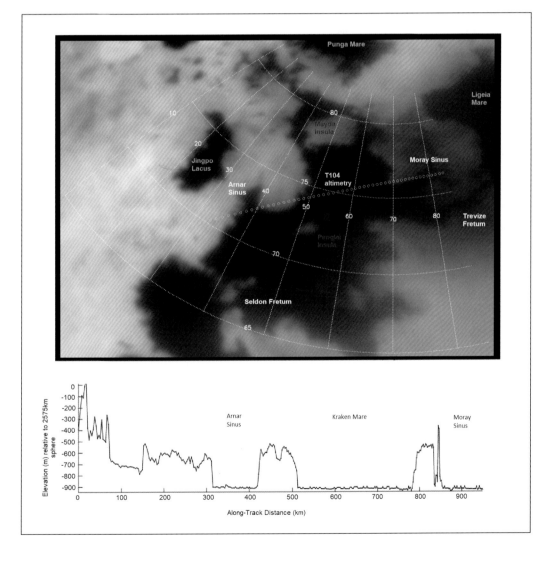

LEFT Titan for climbers. Although Titan's topography is overall rather subdued, there are some steep slopes on the margins of Kraken Mare, its largest northern sea. *(Author)*

TITAN DATA

Property	Value	Context
Distance from Sun	1.5 billion km (9.5 AU)	
Orbital period around Sun (Year)	29.5 Earth years	
Obliquity to Sun	26.5°	Earth 23.4°
Orbital eccentricity around Sun	0.09	Earth 0.017. Mars 0.093
Size viewed from Earth	~0.9 arcsec	Uranus 3.3–4.1 arcsec
Astronomical magnitude	~8.2	Uranus 5.38–6.03
Orbit radius around Saturn	1,220,000 km	Moon is 400,000 km from Earth
Orbital eccentricity around Saturn	0.0288	
Orbital period	15.945 Earth days	
Rotation period	(same, synchronous)	
Obliquity to orbit around Saturn	0.3°	
Mean radius	2,574.7 km	0.4 x Earth
Polar radius	2,574.3 km	
Surface area	83 million km^2	1/6 x Earth
Mass	1.345×10^{23} kg	0.0225 x Earth
Mean density	1,880 kg/m^3	Ice is 900 kg/m^3, typical rocks 2,700 kg/m^3
Surface gravity	1.35 m/s^2	1/7 Earth g. Approx same as the Moon (1/6g)
Escape velocity	2.639 km/s	Earth is 11 km/s
Surface pressure	146.7 kPa	1.5 x Earth's atmosphere
Surface temperature	94 K (−179°C)	
Speed of sound	195 m/s	340 m/s
Light level at noon	~1 W/m^2	~1,000 W/m^2 on Earth (0.001 W/m^2 in full moonlight)
Atmosphere composition	95% N_2, 1-5% CH_4, 0.1% H_2	(Earth 79% N_2, 20% O_2, ~1% Ar, 0-3% H_2O)
Typical surface wind	~0.5 m/s	~5 m/s

One astronomical unit (AU) is the average distance between Earth and the Sun, or approximately 150 million km.

LEFT Titan for aeronauts. The Dragonfly mission will use a lander with a set of eight rotors designed to exploit the dense atmosphere and low gravity to enable it to fly to different locations in Titan's dunefields and an impact crater. *(APL)*

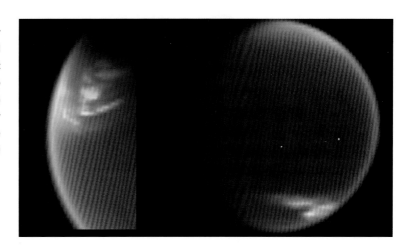

RIGHT Titan for meteorologists. False-colour images of Titan by Cassini from May 2008 and December 2009 showing clouds in yellow. It combines 2µm light (coloured red) that is able to reach Titan's surface, 2.11µm (coloured green) for the lowest part of the moon's atmosphere, or troposphere, and 2.21µm (coloured blue) for the hazy stratosphere. *(U. Arizona/VIMS)*

TITAN CALENDAR

Date	Observation/Event	L_s (°)	Dist (AU)	SSLat (°)
Sep 1979	Pioneer 11 encounter	354		−2.9
Feb 1980	Vernal equinox	360	9.4	0
Nov 1980	Voyager 1 encounter	8	9.5	+4.1
Aug 1981	Voyager 2 encounter	16	9.6	+8.0
Nov 1987	Northern summer solstice	90	10.0	+26.7
Jul 1989	28 Sgr stellar occultation	109	10.0	+25.4
Oct 1990	Author, aged 21, starts work on Huygens	124	10.0	+22.6
Nov 1995	Northern autumnal equinox	180	9.6	0
Oct 1997	Cassini launch	201	9.4	−10.3
Oct 2002	Southern summer solstice	270	9.0	−26.7
Jul 2003	Perihelion	281	9.0	−26.3
Apr 2004	Cassini approach science	292	9.0	−25.0
Oct 2004	First close Cassini flyby (Ta)	300	9.1	−23.5
Jan 2005	Huygens descent	303	9.1	−22.8
Jul 2006	Northern lakes observed by radar (T16)	323	9.1	−16.6
May 2008	Last flyby of nominal Cassini tour (T44)	346	9.3	−6.9
Aug 2009	Northern spring equinox	360	9.4	0
Jun 2010	Cassini equinox mission ends (post T70)	9	9.5	+4.3
May 2013	Ligeia depth sounding (T91)	42	9.8	+18.6
May 2017	Northern summer solstice	90	10.0	+26.7
Sep 2017	Cassini end of mission (post T126)	94	10.0	+26.7
Apr 2018	Aphelion	101	10.0	+26.3
Mar 2021	JWST launch (planned)	135	10.0	+19.5
May 2025	Northern autumn equinox	180	9.6	0
Apr 2026	Dragonfly launch (planned)	191	9.5	−5.2
Apr 2032	Southern summer solstice	270	9.0	−26.7
Nov 2032	Perihelion	279	9.0	−26.4
Dec 2034	Dragonfly arrival (planned)	310	9.1	−21.0
Oct 2037	Dragonfly nominal mission complete	346	9.3	−7.0
Aug 2039	Author age 70 (planned)	6	9.5	+3.2

L_s, the solar longitude, is a way of dividing the year into seasons of equal angular length (90 degrees) when orbital eccentricity makes them of unequal duration. Zero is defined as northern spring equinox, when the Sun crosses the equator northwards (SubSolar Latitude SSLat=0 and increasing). The table shows the Sun–Saturn distance, not meaningfully different from the Sun–Titan distance: the Earth–Titan distance varies by +/-1 AU around the Sun–Saturn distance due to the Earth's movement around the Sun. 1 AU is the mean Earth–Sun distance of 150 million km.

Chapter Two

Titan before Cassini-Huygens

From its discovery in the 17th century to the end of the 20th, Titan gave up few of its secrets. Its exceptional atmosphere motivated the scheduling of a close encounter by Voyager 1, and then the tantalising hints that Titan might have seas set the stage for Cassini-Huygens.

OPPOSITE A Voyager 1 view of Titan (this example in blue light, taken from almost 2 million km away in November 1980) showing the difference in hemispheric brightness and a dark 'hood' over the north polar region. The black dots are 'reseau' marks superimposed on the vidicon detector tube to facilitate geometric correction of the images. There is also some slight vignetting of the corners of the image. *(NASA)*

RIGHT Christiaan Huygens' discovery of Titan on 25 March 1655, around 8pm. *(C. Huygens)*

ABOVE Soon after discovering the satellite, Christiaan Huygens determined its orbital period to be 16 days. *(C. Huygens)*

BELOW Christiaan Huygens' sketch shows how Saturn's appearance from Earth changes due to the fixed orientation of Saturn's rings and pole in space, and thus how Saturn (and Titan) have seasons. *(C. Huygens)*

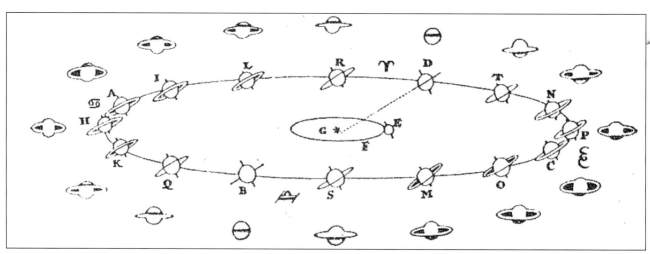

As the most distant planet easily visible to the naked eye, Saturn has been known since ancient times as a point of light in the night sky. The invention of the first blurry telescopes in the early 17th century revealed it to be special – something more than just an orb. The Dutch astronomer, Christiaan Huygens, using a better telescope in the mid-1650s, discerned the distinctive feature of this world – it has rings. Huygens also found that Saturn had a giant moon that would later be named Titan, and that Titan, the rings, and Saturn's equator are all inclined to Saturn's orbit around the Sun in a manner that gives this system seasons similar to those of Earth. But being so far from the Sun – ten times farther than Earth – these seasons would be long. Saturn takes nearly 30 Earth years to travel once around the Sun. And, Huygens realised, it would be very cold – if there was rain out there, it would have to be rain of something other than water.

Among Huygens' contemporaries was the Italian-born astronomer Giovanni Domenico (later Jean-Dominique) Cassini, who established the Paris Observatory. He found that Saturn's rings were not continuous, but there was a gap (which now bears his name). He also discovered several other moons orbiting the planet.

For centuries little more was learned about Titan. Then in 1907 an astronomer's eyeball drew (literally, the observation was recorded as a sketch) further attention to the Saturnian system, when Josep Comas Solà in Barcelona discerned that Titan's tiny disc did not have a hard edge like our Moon, but was progressively dimmed, suggesting that unlike any other moon in the solar system, it had an atmosphere.

Another Dutchman, Gerard Kuiper, working in America, showed in 1944 that indeed sunlight was not reflected from Titan equally at all wavelengths; in particular, specific bands of red and infrared light were absorbed by the moon with the spectral signature of methane gas. Then, as planetary exploration by robotic spacecraft got going in earnest in the 1970s, it was realised that the atmosphere was hazy and that sunlight would slowly destroy the methane, thereby making other organic compounds. The astrobiologist Carl Sagan, who had studied under Kuiper, noted that this process, which may well have occurred on the early Earth, might be the first stepping-stone in chemical evolution, making the building blocks of life. Thus Titan was set as a priority target for exploration.

As planetary science as a discipline really got going around 1970, it proved difficult to measure how warm Titan might be by using a telescope to measure the thermal infrared heat radiation. The problem was that measurements at different wavelengths gave widely different answers. A measurement at 20μm in 1971 by a 2.2m telescope in Hawaii suggested 94K (−179°C), but soon other measurements in the 8–13μm range suggested a brightness temperature as high as 160K. Evidently, gases in Titan's atmosphere were causing emission from different pressure levels at different temperatures.

One possibility was a hot stratosphere over a cool surface, with the air warmed by something that absorbed sunlight – much as ozone absorbs sunlight on Earth to warm our stratosphere. One appealingly self-consistent idea was that some sort of dark aerosol could fulfil this role. A suspended 'dust' might also explain why Titan did not appear blue. Gas molecules, being tiny, preferentially scatter short-wavelength (blue) light. If the atmosphere weren't hazy it ought to be blue, but it was known to be reddish.

Yet another idea could explain the observations so far, and this held out the tempting prospect of a warm oasis in the

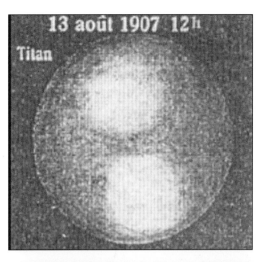

LEFT Josep Comas Solà's sketch showing the darkened edge of Titan's disc, the first indication of an atmosphere. The dark equatorial band was perhaps spurious. *(J. Solà)*.

BELOW Gerard Kuiper's spectrum of Titan can now be reproduced in a matter of minutes with an amateur telescope. *(Author)*

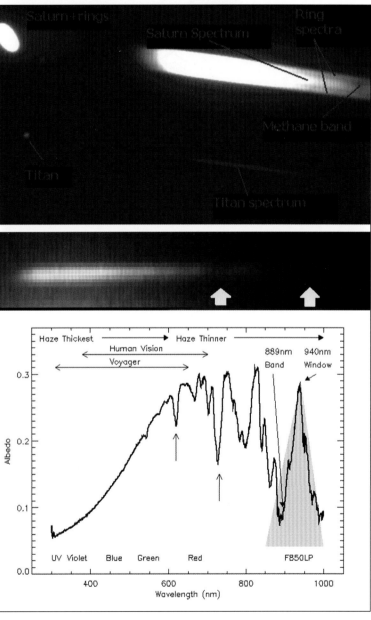

LEFT **This image of Titan just beneath its parent planet was built up slowly as the rapidly spinning Pioneer 11 spacecraft swept its narrow photometer beam across the system.** *(NASA)*

outer solar system. If Titan's atmosphere were sufficiently thick, then it would provide a strong greenhouse effect. There were some early indications that hydrogen as well as methane was present, both of which are greenhouse gases. Perhaps the strong 8–13µm emission was indicating a warm surface. This possibility was greatly favoured by Sagan who, ever-optimistic that a greenhouse from lots of hydrogen might give comfortable conditions at the surface, suggested that surface temperatures as high as 200K were not excluded.

This set the stage for planning the Voyager mission, which featured two identical spacecraft to explore Jupiter and Saturn. Particular tasks for Voyager 1 while on passage through the Saturnian system were to establish the thickness of Titan's atmosphere and the temperature of its surface. Solving this puzzle required a special encounter geometry. From our vantage point, Voyager 1 would have to pass behind Titan and beam a radio signal through the atmosphere. This 'radio occultation' technique, which had been used in the 1960s to determine the thickness of the Martian atmosphere, measures how much the radio ray is bent (refracted) by the atmosphere, allowing its density to be inferred. In combination with data on composition, this allows us to estimate the temperature profile of the atmosphere.

But while the Voyagers were in transit, a lightly instrumented pathfinder, Pioneer 11, made a brief reconnaissance of the Saturnian system in 1979, measuring radiation levels and assessing the hazard of impacts with particles near the rings as it shot past the planet. Pioneer 11 didn't have a camera but its photopolarimeter, a light sensor in a tube that was swept around by the spacecraft's spin, did make some important optical measurements that would help to identify the characteristics of Titan's haze, including the fact that the northern hemisphere of Titan was darker in blue light than the south.

> Since Titan is the easiest body with an atmosphere [for a spacecraft] to enter in the outer solar system, the exploration of Titan may be the primary stage in the study of the organic chemistry of the entire outer solar system.
>
> **Carl Sagan**, *The Atmosphere of Titan*, 1973

BELOW **The spin-scan photopolarimeter on Pioneer 11 gave the first resolved spatial information on Titan's disc since Comas Solà.** *(Author)*

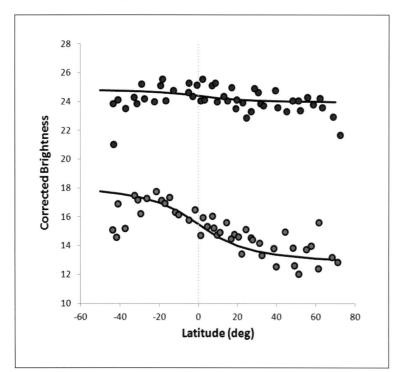

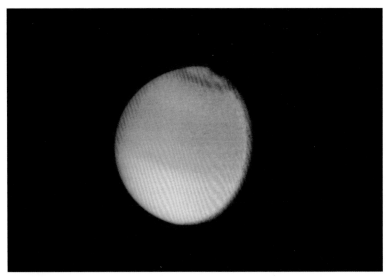

Voyager encounters

The result of the Voyager 1 encounter was interesting indeed. Although the surface temperature of 94K was less than Sagan had hoped, it did establish that there was a significant greenhouse, and allowed the intriguing possibility that methane might exist as a liquid on the surface. The density of the atmosphere at the surface was four times that of Earth, with a pressure of about 1.5 bar. But what landforms might exist at the surface remained unknown. Although Voyager 1 took ~700 images of Titan (and Voyager 2 about 200) the atmosphere proved to be laden with a thick, almost uniform haze (the exception being that the northern hemisphere was fractionally darker than the south) which revealed no details to Voyager's cameras.

Voyager's infrared spectrometer showed evidence of a dozen organic compounds, including propane, acetylene and cyanogen, as well as carbon dioxide, and its ultraviolet spectrometer showed the presence of molecular nitrogen. Thus the atmosphere was a rich soup of organic chemistry.

These hydrocarbons and nitriles were partly made from the fragments of methane molecules, broken up by ultraviolet light from the Sun (and to a lesser extent, from energetic particles in Saturn's magnetosphere). But this process was destroying the methane, and now that we had determined the thickness of the atmosphere it was evident that the amount of methane vapour we can see (which, if it

ABOVE LEFT This close-up view of the limb of Titan by Voyager's wide-angle camera and a violet filter shows rich structure in the haze layers. *(NASA)*

ABOVE This colour composite view from Voyager 2 (with reseau marks removed by interpolation) found the northern hemisphere was still darker than the south, and the north polar hood had narrowed. *(NASA)*

BELOW Although Voyager 2 did not fly close to Titan, its trajectory gave this outstanding high-phase view of the nightside, showing light scattered by the haze all the way around Titan's limb, observed with the clear filter from a range of some 908,000km, three-quarters of the Titan–Saturn distance. *(NASA)*

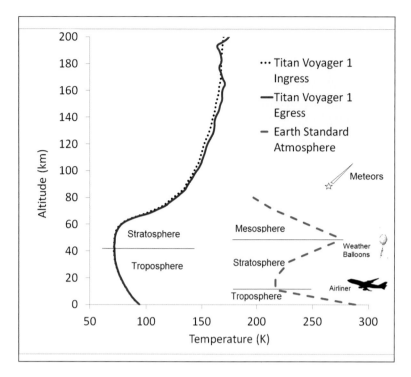

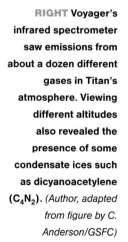

ABOVE The Voyager 1 radio occultation experiment gave profiles of Titan's atmosphere, and showed that the surface pressure was some 1.5 times higher than Earth's. The profile was remarkably Earth-like, with a 'cool' troposphere overlain by a warm stratosphere. *(Author)*

RIGHT Voyager's infrared spectrometer saw emissions from about a dozen different gases in Titan's atmosphere. Viewing different altitudes also revealed the presence of some condensate ices such as dicyanoacetylene (C_4N_2). *(Author, adapted from figure by C. Anderson/GSFC)*

idea offered the prospect that the main product would be ethane, which is also able to exist as a liquid at Titan's surface temperatures, suggesting there might be seas of a methane-ethane mixture.

The next steps would be to use a radar to penetrate the haze and map Titan's surface (as was being planned by NASA for perpetually cloud-enshrouded Venus) and to parachute a probe down through the atmosphere to further investigate its composition. These tasks were to be taken by NASA in partnership with the European Space Agency, which was cutting its solar system exploration teeth with a mission to Halley's comet.

In 1990, while international support for this mammoth project, named 'Cassini', was building, more information was obtained in the form of a radar echo from Titan's surface of several-hour transmissions from the giant 70m radio antenna in Goldstone, California. Titan and Saturn's rings were at the time (and remain still) the most distant radar echoes ever obtained, and while interpretation of the weak reflection was challenging, it seemed to indicate that at least part of Titan's surface was solid.

Other astronomical observations, specifically of starlight being bent by Titan's atmosphere as that body fortuitously passed in front of (occulted) 28 Sagittarii in July 1989, gave credence to the strong 'super-rotating' east–west (zonal) winds that were indirectly

were all to condense onto the surface, would form about 10m worth of liquid) would be destroyed at current rates in only ~10 million years; a tiny fraction of the solar system's 4.5 billion year age. So were we seeing some unusual episode in Titan's history when there had just been a belch of methane, or was methane being continuously supplied? One idea was that there might be seas of methane on the surface that could slowly evaporate to replenish the atmospheric methane that was being converted into heavier organics. This

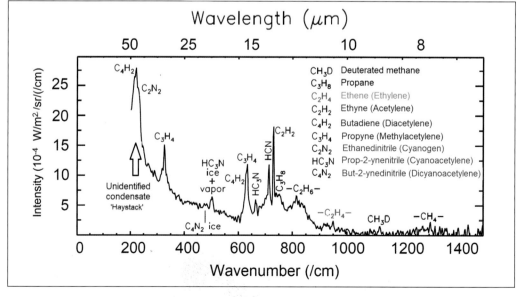

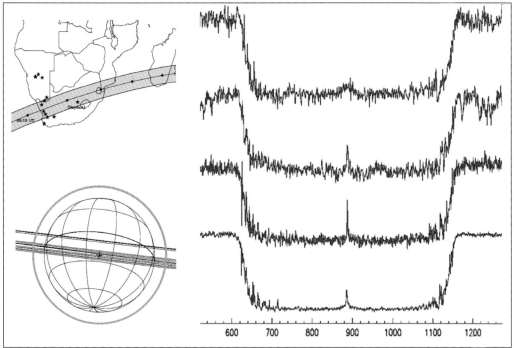

LEFT In a stellar occultation, Titan blocks the direct light from a star (here TYC 1343-1615-1 in November 2003). The blurred edge of the shadow is diagnostic of the atmospheric structure. Sometimes, when the star is directly behind Titan, its light can be bent around the obstacle by the moon's atmosphere acting like a lens, creating a central flash. The technique is especially powerful when observers, equipped even only with 'amateur' 12-inch telescopes, deploy in a line to observe lightcurves simultaneously, drawing multiple chords across Titan's disc) to measure the 'shape' of the atmosphere. *(Author, with material from B. Sicardy)*

inferred from Voyager thermal infrared measurements of atmospheric temperatures. Winds would be important in estimating how far an entry probe would drift during its long parachute descent, and in turn might influence where a relay antenna on Cassini would have to point to receive the probe's signal. Unfortunately, neither the Voyager data nor the stellar occultation indicated which direction, eastward or westward, the probe would drift.

One other important scientific result around this time was the realisation that although Titan's hazy atmosphere was opaque to visible light, and therefore to Voyager's cameras, it would be possible to penetrate the haze at selected wavelengths in the near-infrared (such as at 940nm, which is widely used in TV remote controls). As a result, not only Cassini's radar, but also its near-infrared spectrometer and camera would be able to sense the surface.

Similar select 'windows' in the near-infrared also reveal Titan's surface to scrutiny. While the methane absorptions that heralded the presence of an atmosphere to Kuiper block much of the infrared spectrum, and the thick tholin haze blocks blue and green light, about 10% of the light filters through the haze in about half a dozen windows located between methane absorptions, and the amount of light that is reflected by Titan at these wavelengths depends to some extent on how reflective the ground is.

RIGHT The visible and near-infrared spectrum of Titan. The diamonds show calculations of how bright Titan should be, according to laboratory measurements of organic haze reflectivity, distributed in a methane-laden atmosphere. The line shows the spectrum of Titan measured by a terrestrial telescope. *(Author)*

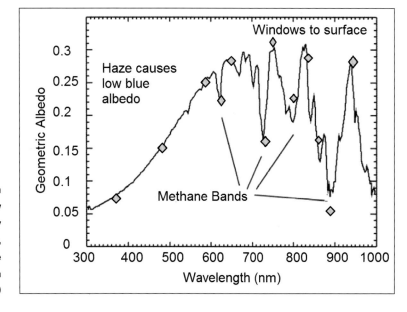

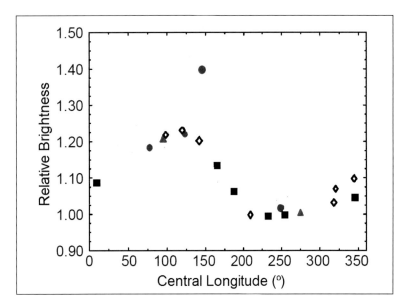

ABOVE Titan's 2μm lightcurve. Different observations from different telescopes (shown by symbols/colours) line up in a consistent pattern, indicating that Titan's leading hemisphere (90° longitude) is brighter than the rest. A rogue point of exceptional brightness at a longitude of 140° was a cloud system that popped up in 1995, one of the first indications of active weather on Titan. *(Author)*

BELOW Although the entire near-infrared spectrum is rich with compositional information (the reflectances of different outer solar system ices are shown offset for clarity), Titan's methane allows only narrow windows (grey) of light through, making it difficult to identify Titan materials from space. *(Author)*

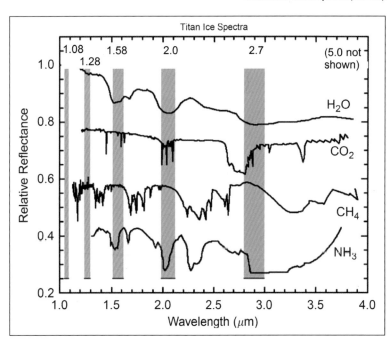

First glimpses of the surface

Because Titan rotates synchronously in a tidally locked manner (it always presents the same face to Saturn, just as our Moon does to us), its place in the orbit corresponds directly to the longitude of the point beneath a distant observer. Hence, when Titan appears furthest from Saturn (its greatest eastern elongation) the disc, as seen from Earth, is centred on 90°W. The prime meridian (zero longitude) is defined as the centre of the sub-Saturn hemisphere. If an observer measures the brightness over time and plots it versus fractions of an orbit ('phase') then the graph, called a 'lightcurve', is effectively one of reflectivity versus longitude; a one-dimensional map. The first efforts to make such lightcurves in the early 1990s showed that Titan's surface was not uniform. In particular, the leading hemisphere of Titan (90°W) was consistently brighter than the trailing hemisphere. It was therefore possible to infer that the moon was not covered by a global ocean, but little else could be said.

In 1994 the Hubble Space Telescope, observing Titan not only at 940nm but also in visible red light, made the first maps of bright and dark areas. In addition to the near-infrared maps, the Hubble images at visible wavelengths showed Titan's atmosphere to be 'upside down' relative to what Voyager had seen in 1980. This indicated a seasonal shift in the amount of haze in the two hemispheres. For the Voyagers in 1980/81, it was the northern spring equinox and the north was about 20% darker in blue and green light. Now it was half a Titan year later. The ability of Hubble to image in light that was absorbed by methane (and hence only showed high-altitude haze above the methane-rich troposphere) established that the difference was in the amount of haze, and suggested the haze was being blown by north–south (meridional) winds in the global circulation pattern.

Then in 1998 ground-based spectrometry revealed that Titan had clouds. Not the global-scale, high-altitude, almost uniform haze, but smaller, rapidly changing clouds in the troposphere which occupied only about 1% of Titan's disc. This implied methane clouds, and

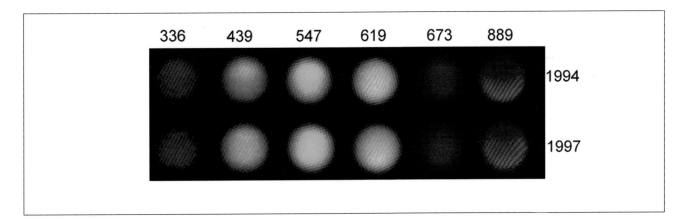

ABOVE Titan images from the Hubble Space Telescope. Even only 20 pixels across, it is obvious that ultraviolet (336nm) and methane band (889nm) light probes only the highest levels of the atmosphere and so the optical diameter is larger at these wavelengths. The northern hemisphere is bright at blue (439nm), the opposite of what Voyager had reported half a Titan year earlier. The 'smile' in the methane band at 889nm shows the haze to be more abundant in the southern hemisphere. *(Author)*

RIGHT Year-to-year variations in Titan's haze shown in a series of HST images (the changing seasonal geometry is shown at left). The UV (336nm) and red (673nm) appearance stays somewhat constant, while the blue (439nm) north–south asymmetry declines. The 889nm methane band asymmetry reverses altogether with the 'smile' becoming a frown. *(Author)*

RIGHT (Left) an image of Titan by the Hubble Space Telescope in 1994, showing near-infrared light longward of 850nm, including that at 940nm which sneaks through the haze between methane absorption bands. The lower limb is bright owing to thicker haze at that latitude and season, but irregular dark markings (about 10% of the total light) show bright and dark regions on the surface. Only features bigger than about 10% of Titan's diameter are distinguishable, so their nature could only be guessed at. At right is an ultraviolet image from 2002, showing a dark hood over the south pole. *(STScI/Author)*.

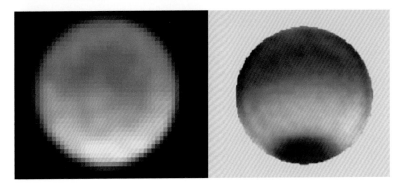

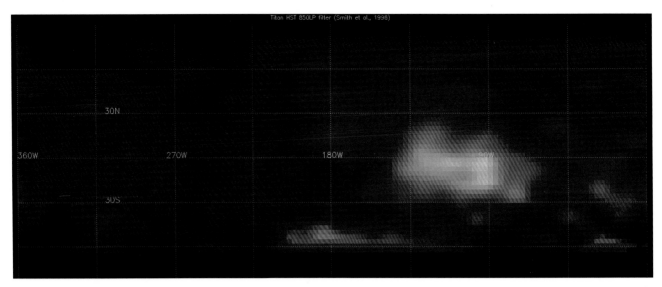

ABOVE A set of 14 images of Titan acquired in 1994 were assembled into a map of surface contrasts at a wavelength of 940nm. Since the images were not evenly spaced in time, there was a bit of a gap close to the sub-Saturn point (0° longitude) and so the map was plotted with 180° in the middle in order to de-emphasise the ugly gap. Most Cassini Imaging Science Subsystem (ISS) maps at that wavelength, and the radar maps which used data from the Hubble Space Telescope in early planning, employed that same convention, although near-infrared maps by the Cassini VIMS instrument tended to use a more imperial zero-longitude-at-the-centre style. *(STScI/U.Arizona)*

perhaps an active hydrological cycle. Despite much speculation, the paucity of evidence of any clouds in the Voyager data had led some people to argue that discrete clouds and convection might not happen. These new observations were supported by some tentative indications of a transient bright region in Hubble images.

Unfortunately, Hubble observing time was deemed too precious to devote to monitoring for clouds which might or might not appear, so more systematic studies of Titan's clouds did not really begin until around 2001, when the large telescopes with 'adaptive optics' systems (fast-adjusting mirrors to compensate for the shimmering in our atmosphere) were able to discern dramatic cloud systems at the south pole, where it was now mid-summer on Titan. As these systems improved, clouds started to be observed frequently, with patterns which hinted that Titan's geography or topography might favour certain locations.

Between them, the Hubble images and the near-infrared maps from ground-based telescopes indicated the presence of bright and dark regions on Titan's surface, but only with a level of detail resembling a naked-eye view of our Moon. The leading-face brightness was found to be an Australia-sized region (subsequently named Xanadu) and it was speculated that this was a mountainous highland region, perhaps washed clean by the methane rainfall (the logic being why would it stay bright if the muddy organic products of methane photolysis drizzled from the atmosphere everywhere). And perhaps the dark regions were seas.

Remarkable radar observations were brought to bear on this question. This was a collaborative effort in which the Goldstone Deep Space Network antenna transmitted a signal and the Very Large Array at Socorro in New Mexico received the echo. In this way, Titan's radar reflectivity was measured in 1990 and again several years subsequently. These challenging observations showed the surface to be too broadly reflective to be entirely covered in seas. Furthermore, the leading face was brighter than the rest. But again, interpretation was difficult.

More quantitative and detailed observations came in 2002-03 when Titan was investigated by the new powerful radar transmitter of the 305m Arecibo radio telescope in Puerto Rico. In fact, the echo was sufficiently strong to divide the signal into different frequencies, Doppler-shifted by Titan's rotation. Overall, the echo implied a scattering surface that was either rough at the wavelength scale of several centimetres or included some reflective proportion in the shallow subsurface, but on a few occasions there was a prominent spike

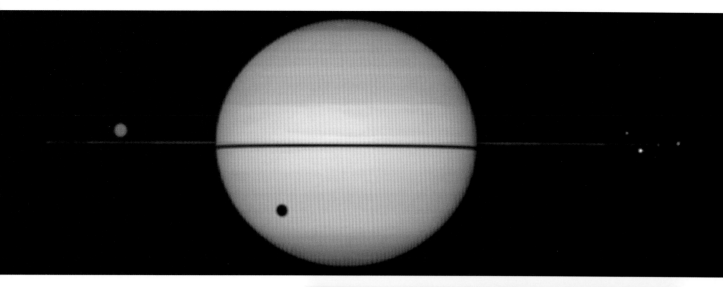

ABOVE During the equinox and ring-plane crossing in 1995 the rings were edge-on, and Titan could cast a shadow on (and be eclipsed by) Saturn. This Hubble Space Telescope image shows Titan and its shadow, together with several smaller moons. In fact, close inspection of the size and shape of the shadow at different wavelengths showed that the structure of Titan's haze varied with latitude. *(STScI)*

in intensity right at the zero-Doppler centre of the echo which was suggestive of some very flat patches, and the simplest explanation was that these were lakes or seas of liquid hydrocarbons. Cassini, with its radar and infrared cameras, would find out for sure.

The site for the Huygens probe's entry and descent was determined by considerations of the necessary illumination, entry angle (affecting the required heat shield performance) and the communications-relay

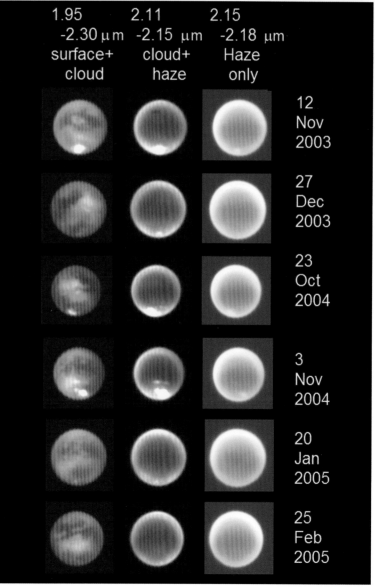

RIGHT Titan viewed by the giant Keck II telescope with adaptive optics beginning in the early 2000s. On the left are images at 2.0μm (little affected by methane absorption) showing Titan's surface and troposphere. The middle images at 2.12μm block the surface and emphasises the troposphere. There are prominent discrete clouds at southern high latitudes. In the column, at 2.16μm, methane absorbs light from the lower atmosphere, thus it shows only haze in the stratosphere, which by that time was more abundant in the north. *(Author, from W. M. Keck Observatory/M. Brown et al.)*

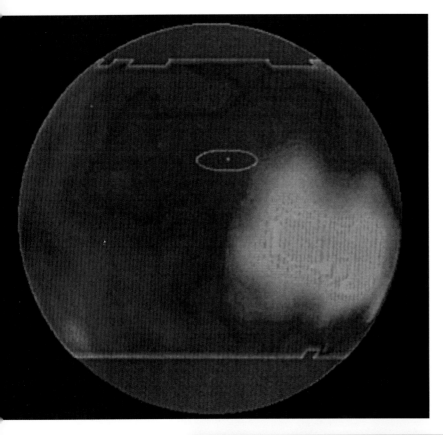

LEFT The original descent location for Huygens (white ellipse) was at 9°N and 145°W, which turned out to be just off the northwest edge of Xanadu (the bright yellow region on this Hubble Space Telescope map, projected onto a sphere). *(Author)*

geometry with the Cassini spacecraft as that flew by. These factors put the descent location at just north of the equator. At the time that the target was chosen (it was not strictly a 'landing site', since there was no guarantee that the probe would survive contact with a completely unknown surface) the observations had yet to be made that would place it just off the northwest edge of the Xanadu continent.

However, a problem with the Huygens telecommunication system that was discovered only in 1999, with Cassini in deep space, necessitated a redesign of Cassini's trajectory and, with it, the entry date and location for Huygens. The new site would be somewhat further to the west and at 10°S. This was again near the border between a bright area and a

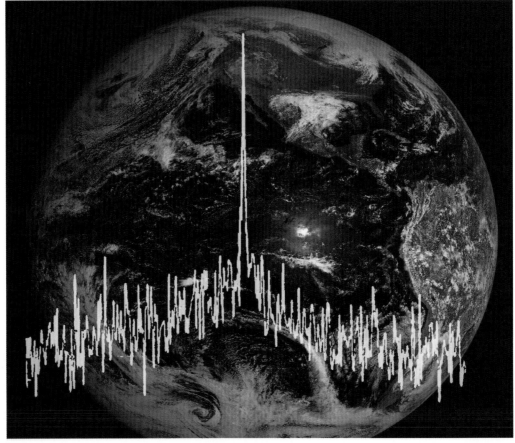

RIGHT Reflections on two complex worlds. The background optical image of Earth was taken by a geostationary meteorological satellite, and shows clouds with a higher albedo than land, which in turn is brighter than the ocean, except for the striking specular reflection of the Sun just to the right of centre. The curve is a radar spectrum of Titan taken by the Arecibo radio telescope, and is a one-dimensional equivalent of the Earth image. The sharp spike indicates a strong specular reflection from a smooth surface on Titan, hypothesised at the time to be a lake or sea. *(Author)*

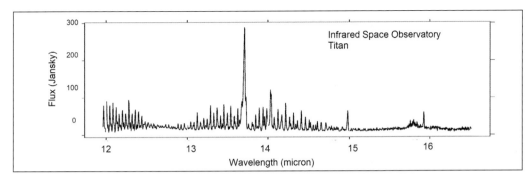

LEFT The Infrared Space Observatory of the European Space Agency observed Titan at a much higher infrared resolution than Voyager, and provided a hint of the rich possibilities of Cassini's CIRS (Composite InfraRed Spectrometer) instrument. (Author)

dark area, but the mystery remained as to what that would mean in terms of terrain.

One of the major uncertainties for Huygens was whether the winds of Titan would carry the probe to the west or to the east during its 2.25hr descent. Strenuous efforts were made using telescopes to try to resolve this question. There were faint hints in Voyager and Hubble data that features which might be clouds travelled in the expected prograde (forward, i.e. in the direction of Titan's rotation, west to east) direction, but these were not definitive. In the early 2000s visible, infrared, and millimetre-wave observations by the largest telescopes provided high-resolution spectroscopy to measure the Doppler shift at the eastern and western sides of the moon's disc, confirming that the winds were prograde. But by then the mission redesign meant that the signal from the probe would be receivable either way.

BELOW Two dishes of the Institut de Radioastronomie Millimétrique (IRAM) telescope array in France, seen in 1993 with the author for scale. They work at shorter wavelengths than 'classical' radio telescopes, and so the finish of the parabolic reflectors has to be rather mirror-like. These telescopes were used to measure the abundance of carbon monoxide and nitrile gases in Titan's atmosphere and also to estimate the winds. (Author)

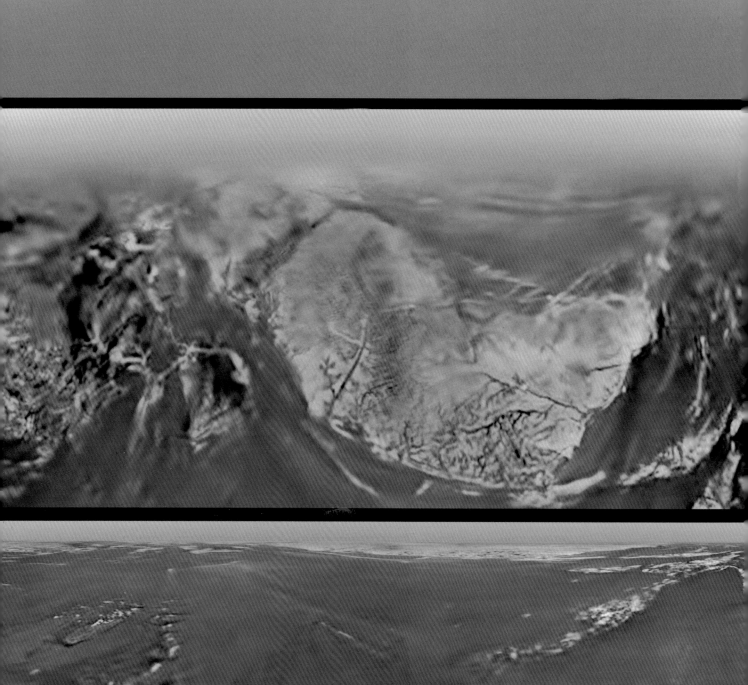

Chapter Three

Titan unveiled

After a billion-mile voyage lasting seven years, Cassini arrived at Saturn in 2004 for an epic survey of that system. In addition delivering the Huygens probe to Titan in January 2005 for the most distant touchdown ever made on a planetary body, the orbiter performed over 100 close flybys of the moon.

OPPOSITE The Huygens imagery, coloured using spectral data, projected as panoramic Mercator views from altitudes of about 100km, 10km and 0.4km. *(ESA/NASA/JPL/U. Arizona)*

After its launch in October 1997, Cassini's trajectory utilised the 'gravity assist' or 'slingshot' effect of no fewer than four planetary flybys. First it headed inward toward Venus, whose gravity would deflect it out beyond Mars for a deep space manoeuvre using its main engine that would aim it back at Venus for another boost. The second Venus encounter, in summer 1999, produced an Earth flyby a few months later that gave the craft the kick it needed to arc out to Jupiter, which in turn sent it to Saturn. There were some limited science observations during these encounters, but the opportunities were constrained at Venus and Earth/Moon by the need to maintain the high-gain antenna dish pointed at the Sun to shade the body of the spacecraft (whose thermal design was optimised for the cold of the Saturnian system, where sunlight was 100 times less intense). Some of the instruments were not fully operational (e.g. their covers were not deployed), and software to create observing sequences such as image mosaicking was not developed until the Jupiter flyby in 2000.

With its array of a dozen formidable instruments, Cassini was equipped for a wide-ranging investigation of the entire Saturn system but one of the main foci of attention was Titan, and several instruments were designed specifically with it in mind. The Huygens probe was far better instrumented than its predecessor on the Galileo mission to Jupiter, and the amount of data that it was to send back would be considerably greater.

Cassini's instruments included a radar, principally for mapping Titan. Its camera and near-infrared imaging spectrometer would also be able to peer through haze to study the moon's surface. These, and the other spectrometers, would mainly study Titan's atmosphere and its interaction with the Saturn environment.

The radar transmitted bursts of pulses, each typically swept in frequency ('chirp') to facilitate processing of the echoes to maximise the signal without ambiguities; basically the processing isolated the reflected energy by its time-of-flight (and hence its range from the spacecraft) and its Doppler shift (which effectively measured the angle away from the spacecraft's trajectory).

When the radar was being operated in its 'scatterometry' mode at long distances, all the echo energy within the main antenna beam was just added up. The beam was one-third

OPPOSITE Cassini released Huygens five months after their arrival at Saturn. Cassini made its first two close flybys of Titan (TA,TB) in October and December 2004 with Huygens attached. Huygens was actually released at Christmas 2004, while Cassini was arcing away from Saturn towards the moon Iapetus. *(ESA)*

BELOW The Huygens probe coasted through cold space for three weeks, before making a hypersonic entry into Titan's atmosphere and a 2.5-hour parachute descent, surviving impact and continuing to operate for over an hour. *(E. Carreau/ESA)*

Final Close Titan Flyby (T-126)
Gateway to the Grand Finale
April 22, 2017, *Earth Day*

ABOVE Extensive planning went into every one of the thousands of observing sequences on Cassini – this graphic shows how the view angle of the radar was adjusted several times during the last Titan encounter, so as to hit specific scientific targets. *(NASA/JPL)*

of a degree wide, so at a range of 50,000km it defined a footprint 250km across – about the same size as a Hubble pixel. Closer in, with the antenna pointed vertically down, the principal information was in the echo as a function of time (only in particular cases was the Doppler shift exploited). But the real magic of radar came when the echo was chopped up in near-orthogonal range and Doppler directions to create a little image (known as Synthetic Aperture Radar, or SAR). These tiny instantaneous images were stacked up as the spacecraft swept past Titan, to create a long, thin image. In Cassini's case such an image would be too narrow to interpret, so the radar was switched in rapid succession between five overlapping beams to make a broader swath which typically spanned 200–300km at closest approach. The resolution could be as high as ~300m.

One feature of radar images is that because they observe a retroreflection, much like taking a flash photograph in a dark room, certain surfaces can glint very strongly; in the jargon of SAR, the radar has few 'looks', so the image appears 'grainy' or 'speckled'.

A trick devised by the radar team a couple of years after arrival at Saturn was to exploit the overlap between the five sub-beams. If the assumed elevation of a given point in the overlap region were incorrect, the feature would appear to have different brightnesses, leading to an ugly 'seam'. By adjusting the assumed height along the swath, the seam could be smoothed over, and a topographic height profile developed. These 'SARtopo' profiles, which covered much more terrain than the altimeter observations, proved very valuable in determining the shape of Titan.

The images from the ISS camera system are easier to describe. Just like HST images, Titan's surface features contributed only a few per cent of the total signal, so by subtracting away a smooth image it was possible to isolate the surface reflection. Since the HST images were all taken with a constant geometry, an average image could be used as the background, but the Cassini ISS images were taken at widely varying distances to Titan, at different angles from vertical, and with different angles of illumination. So simultaneous images were taken with a filter (CB3) that sensed the surface at a wavelength of 938nm and another (MT1) for the weak methane band at 619nm that sensed the haze, and the maps were generated as the ratio of the two images.

Cassini's Visual and Infrared Mapping Spectrometer (VIMS) combined a 1×256 array of indium antimonide (InSb) infrared detectors with a diffraction grating to produce the spectrum of light from a 23cm telescope over the range 0.85–5.2μm of a single pixel of 0.25×0.5 milliradians. This pixel could be scanned in one dimension to yield a slice of 64 spectral pixels or in two dimensions to produce a 64×64 pixel image cube. Most of the Titan mapping was done in this latter way, but a few very close observations (when the spacecraft would sweep past too quickly to scan a cube) were made in 'noodle mode', taking the 1x64 slices in rapid succession along the groundtrack.

One other observation type for VIMS (and Cassini's Ultraviolet Imaging Spectrometer, UVIS) was to observe the light from a star or the Sun as the trajectory swept the line of sight through Titan's atmosphere. These solar and stellar occultations provided profiles of the absorption of different wavelengths of light as a function of altitude, to profile the abundance of a variety of different gases, as well as the haze. These profiles were obtained throughout the mission for a range of latitudes and seasons.

Profiles of atmospheric density (and with some assumptions, temperature) were obtained by using radio telescopes on the Earth to observe the radio signals from Cassini as the craft flew behind Titan. This was the same type of radio occultation experiment originally performed by Voyager to measure how dense the Titan atmosphere was, but Cassini's experiment could be done at several wavelengths and with rather higher precision. Again, these observations were made a number of times in order to study the difference between low and high latitudes, and also how the profiles changed with the seasons.

For some of the occultations, and sometimes performed as a standalone experiment, Cassini aimed its radio beam onto Titan's surface in such a way that the reflection was measured on Earth. This so-called 'bistatic radar' experiment in which energy was bounced at a variety of angles, and at different wavelengths from the radar instrument, was designed to provide complementary information on the surface scattering properties, especially over Titan's seas.

The radio equipment on Cassini was also used in a completely different way to measure very precisely the trajectory of the spacecraft. When Cassini made flybys close to Titan the gravity field would alter its velocity. Provided the encounter was not so close as to induce significant atmospheric drag, these changes in velocity could be measured with sufficient precision to be able to calculate deviations of the field from a spherical geometry. Indeed, not only could the polar flattening of the field be measured, but also the changes in the field as Titan was tidally distorted by Saturn's gravity in its elliptical orbit.

Of course, mission planners had to make difficult choices. For example, the high-gain antenna could not point in two directions at once. It could be pointed at Titan for radar (or bistatic) experiments, or at Earth for gravity (or at the image position of Earth refracted through Titan's atmosphere for a radio occultation). And since the antenna was mounted orthogonally to the optical remote sensing instruments, if the surface was studied by VIMS/ISS it was not possible to use the radar.

Some of Cassini's instruments were insensitive to the attitude of the spacecraft. It was possible to take measurements of the magnetic field in any orientation, and the Radio and Plasma Wave System (RPWS) could detect the crackle of lightning discharges on Saturn without requiring a specific orientation.

The instruments on Cassini for measuring the particle environment at Titan had some pointing freedom, but generally were interested in a different set of directions. For example, the arrival direction of particles was dictated by the direction in which the plasma was flowing (generally being swept by Saturn's magnetic field) and the velocity of the spacecraft relative to Titan, and different instruments measured the mass, charge and energy of the various particles. Negative ions and electrons were measured by the CAssini Plasma Spectrometer (CAPS), while positive ions and neutral gases were measured by the Ion and Neutral Mass Spectrometer (INMS) and particles with very high energies by the Magnetospheric IMaging Instrument (MIMI). At the high mass end, Cassini had a Cosmic Dust Analyser (CDA) principally to study the rings and the material ejected by Enceladus, although it did once make an observation which illustrated how Titan's atmosphere created a 'shadow' in the dust flux.

Titan's atmosphere was studied by UVIS, VIMS, ISS, and a Composite InfraRed Spectrometer (CIRS). Collectively known as the Optical Remote Sensing (ORS) suite, these instruments were co-boresighted on a pallet so that they could observe jointly. CIRS used a telescope to analyse the 5–100μm thermal infrared radiation emitted by Titan's atmosphere at a spectral resolution much higher than Voyager. Both VIMS and CIRS had infrared detectors which had to be kept very cold, so they needed radiators that faced cold space. These radiators could not be pointed at the Sun or even in the forward direction in close Titan flybys (where frictional heating of the atmosphere might warm the radiators and increase the 'noise' on the detectors, or even cause thermal stress). These thermal constraints, combined with the clamour from each instrument's team to point in different directions during every Titan flyby, made the choreography of each encounter very demanding.

A few months before Cassini reached Saturn, the resolution of the ISS narrow-angle camera on Titan exceeded that of Hubble, and thereafter it kept getting better! Orbit insertion of Cassini was on 3 July 2004, when the spacecraft fired its engine to brake into orbit around the ringed planet. Interestingly, this was the closest that the spacecraft would come to the rings until the very end of its mission, some 13 years later.

Although the scientists were distracted with remarkable close observations of Saturn and the rings (as well as the earlier very close look at the outermost Saturnian satellite, Phoebe), they were eager for the observing opportunity called T0 ('T-zero') in which the south polar regions of Titan could be inspected from a distance of ~300,000km as the spacecraft headed out on its highly elliptical initial orbit of the planet.

When it was close to the apoapsis of its orbit, Cassini fired its engine to raise its periapsis clear of the ring system. This also targeted the first close encounter of Titan, called TA. The plan had originally involved T1, T2, etc., with the Huygens probe being delivered on T1, but the orbital tour had been reconfigured to overcome the in-flight discovery of a flaw in the Huygens radio system. To minimise confusion in the later flybys, which were unchanged, the first few flybys were renamed TA, TB and TC (the new flyby for delivering the probe) before the old plan was resumed at T3.

The TA encounter in October 2004 saw Cassini flying just 2,500km above Titan, and gave many of the instruments their first opportunity to study the world up-close. An early priority was to learn the density of Titan's upper atmosphere, since this would determine how low subsequent flybys of Titan would dare go. Lower altitudes would give richer data – up to the point where the atmospheric drag on the spacecraft might overpower its thrusters and deny it the ability to point accurately.

The first flyby showed that although the intrinsic resolution of Cassini's camera was ~10m, the scattering in Titan's thick atmosphere blurred the details of the surface to such an extent that it could not see features smaller than a few kilometres across.

The Voyager radio occultation experiment had given us an appreciation of the overall thickness of Titan's atmosphere, so the size of the heat shield and parachute needed to achieve the desired descent time, starting from an altitude sufficiently high to study the haze profile (~150km), was straightforward to design. But some aspects were uncertain as Huygens was being developed, not least our total ignorance of the nature of the surface!

First was how much methane was present in the atmosphere. This was important because as the probe performed its hypersonic entry, the tortured atmosphere would glow like a meteor trail. In fact, the combination of methane and nitrogen on Titan meant that this glow would be substantially violet in colour, due to a strong carbon-nitrogen emission at 387nm wavelength. The more methane in the atmosphere, the stronger this radiation might be, and the hotter the heat shield would get. The shield was designed conservatively, but the methods to estimate the heat loads were still somewhat in their infancy and the methane (and argon) abundances were not known with certainty. Would the design be conservative enough? And would the winds be as expected?

Huygens descent

Prior to discussing the interpretation of some of the first things seen by Cassini at Titan, it will be useful to illustrate the process of trying to understand a landscape that is at once exotic and familiar. We are fortunate to live on a very complex planet that has examples of processes that seem to be replicated on Titan, albeit probably at very different rates and with rather different working materials. But even guided by terrestrial analogues, many mysteries remain that will likely not be resolved until future missions

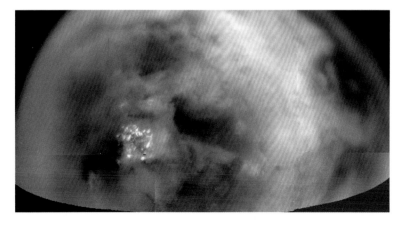

BELOW A mosaic of early (T0) Cassini ISS observations of Titan's south polar region. There is a cluster of bright clouds around the south pole. Some of the dark areas nearby are liquid-filled, but the dark region at lower latitudes (seen at the upper-right here) would prove very different. *(NASA/JPL/SSI)*

provide a closer examination of its surface (or even of its subsurface).

But one area on Titan did receive closer scrutiny than the rest, by the Huygens probe, and its observations instantly laid to rest many debates. Some observations confirmed what had been suspected – for example, the temperature profile measured during the probe's descent plotted almost perfectly on top of the Voyager-derived model. That is not to say the new observation was useless – there was a range of possible models, and the much more detailed and directly measured profile revealed features impossible to detect by radio occultation, notably gravity waves in the stratosphere and kinks in the temperature profile near the surface that indicated the structure of the planetary boundary layer.

The measurements by the probe's optical instrument, the Descent Imager/Spectral Radiometer (DISR), were important in determining the scattering and absorbing properties of the haze as a function of altitude – vital information for the quantitative analysis of VIMS and ISS data. The models had presumed that the haze would be washed out of the atmosphere at low altitudes, but this proved not to be the case – the haze went all the way to the ground, which meant that for much of the probe's descent its impression of the surface was muted; only at an altitude of ~15km did surface contrasts emerge from the noise. The DISR camera had a small aperture and used short exposures to avoid motion blur as the probe was buffeted on its parachute. It used a broad bandpass of red and near-infrared light from ~600–1,000nm, and so was more sensitive to haze scattering than the ISS on Cassini (which had a large telescope and a rock-steady platform for exposures long enough to view using only the 940nm wavelength).

DISR recorded how the haze and methane affected the upwelling and downwelling spectrum, as well as how the light was scattered near the Sun direction (the solar 'aureole'). Although the haze was a heavy overcast without crisp shadows, there was still some topographic shading as different slopes were exposed to more or less of the brighter parts of the sky.

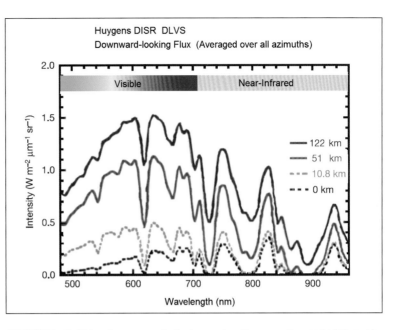

ABOVE The light levels progressively dropped as Huygens descended through the haze, particularly in the methane bands at 619, 729, 793 and 889nm. Essentially no light reaches the surface in the longer bands, nor indeed at blue wavelengths shortward of 500nm. *(Author, from DISR data)*

BELOW Part of the Huygens descent image set. Individual images were only 256 pixels tall and often highly compressed, but the overlap between images allowed a stereo model of surface topography to be constructed and how the probe drifted in the wind to be estimated. *(NASA/ESA/U. Arizona)*

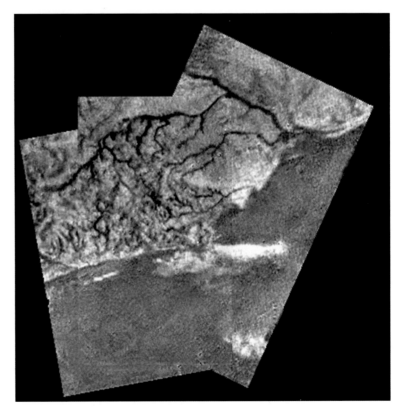

ABOVE Images taken during the Huygens descent were assembled into a map mosaic, and reprojected here to simulate a wide-angle 'airplane window' view from an altitude of 15km. The two dark parallel lines at the upper-right are dunes that would later be located in Cassini's radar data, thereby enabling the two datasets to be matched up. *(E. Karkoschka)*

As the surface hove into view, the debate about whether the surface of Titan was shaped by hydrological processes was immediately resolved in the positive. The wiggly lines present in distant ISS images had been somewhat equivocal, but Huygens revealed a clearly branching network of dark stream channels on a brighter highland, draining into a wider dark area that may have been a prior river or lakebed. There was some debate about the extent to which the channels were dark because of dark material, or dark because of topographic shading (steep valley walls not seeing as much of the bright sky), but there was no doubt that these features were formed by rainfall. Consequently, clouds like those seen puffing up over the south pole must sometimes also have formed at low latitudes and the precipitation did not evaporate in the lower atmosphere but reached the ground. This implied a heavy, transient rainfall. Off in the distance, two dark streaks could just be discerned in the probe descent images. They were instrumental in correlating Huygens' images with data taken by Cassini.

Evidence of the episodic rainfall and surface flow was most apparent in the images returned from the surface. The DISR camera looked outward and downward, its field of view being panned around as the probe rotated under its parachute (as it happened, the opposite way to that planned) and the images from the surface were near-identical, looking out from the knee-high viewpoint in whatever direction the probe happened to be pointing when it skidded to a halt at impact. It saw 'rocks', but unlike the angular fragments that litter Mars and the tabular volcanic slabs on Venus, they were rounded cobbles resting on a darker fine-grained substrate material.

RIGHT A digital elevation model of the highland near the Huygens landing site. The range from blue to red is about 100m. *(Courtesy Chloé Daudon/IPGP)*

Huygens could eliminate the effects of the atmosphere that had so challenged interpretation of VIMS data, in that it could view some surface material through only a short atmospheric path, without significant scattering or absorption. The overall illumination was of course filtered by its passage through the haze and methane, but a tiny patch of ground was lit by a spotlamp on the probe – blasting the surface with light ~200 times stronger than sunshine at noon on Titan. The wavelength range of the DISR spectrometer only extended to 2.5μm, not the 5μm of VIMS, so there is some ambiguity as to whether these 'rocks' were made of water ice or some organic material.

Accelerometers recorded the 15g impact of the probe on the ground, showing the surface to be somewhat cohesive, like damp sand. A protruding force sensor, a penetrometer, supported this interpretation, but it also indicated the presence of a few millimetres of soft dust resting on the uppermost surface. There was also a force spike which one wag ventured was suggestive of a crust like crème brûlée, although it was more likely to have been simple contact with a pebble or cobble.

The dampness of the ground was indicated by a heated inlet on another Huygens instrument, the Gas Chromatograph/Mass Spectrometer (GCMS). It recorded methane levels rising after impact as the inlet sweated moisture out of the ground, as well as traces of ethane, acetylene, and carbon dioxide. Even basic housekeeping data from the probe was brought to bear – heat was wicked away from the inlet more than would be the case for a dry sand (like wet sand at the beach 'feels' colder to one's finger).

Because it had not been known whether Huygens would survive contact with the surface, or how long it might last if it did, there was no post-landing operating sequence; it simply kept repeating essentially the same measurement sequence. And so several hundred images were taken that differed for the most part only by occasional artefacts due to the algorithm which compressed the data interacting with different exposure times, and a few 'cosmic ray' tracks that may well have been neutrons from the radioisotope heaters that had warmed the probe during its coast through space to Titan, rather than cosmic rays. But one image had a blob in one corner that could not be explained this way. One physical scenario that fits the observed characteristics was a 4.5mm drop of methane a few centimetres in front of the lens aperture, possibly produced by moisture cooked from the ground by the spotlamp condensing on the cold camera baffle. If so, then many dewdrops had likely formed on the baffle and we were lucky that one dripped off just when the camera was taking an image. The body of the probe was insulated by a thick layer of foam, so the ground beneath would not have been affected much by heat leaking from the warm interior of the probe.

Although the terrain at the landing site itself was rather flat, close study of the descent images revealed steep gullies just a few hundred metres away. Another set of channels had a different character from those on the highland. Rather than winding and branching at acute angles from some invisibly tapered origin, these were straight, branching at right angles

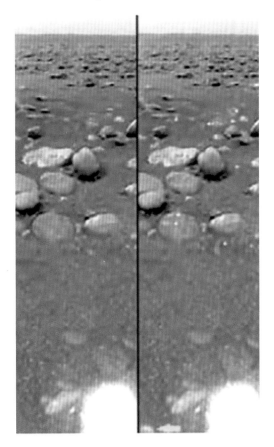

LEFT Close examination of the Huygens landed images showed a small blob at the bottom (right image, arrowed) that was not present in the others (left image). Methane was sweated out of the ground by the surface science lamp and evidently condensed on the cold metal. The blob may have been a drop in the act of dripping off the camera. *(E. Karkoschka/U. Arizona)*

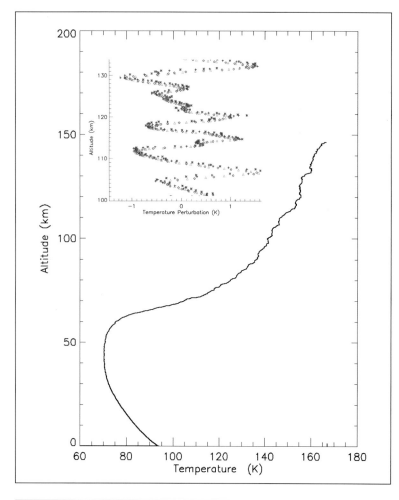

LEFT Gravity waves in the stratosphere of Titan are difficult to pick out in the raw data from the Huygens probe, because the scale of the temperature profile plot is usually set by the large (~100K) difference in stratospheric and tropopause temperatures. Subtracting a smoothed temperature curve leaves a residual profile which shows magnificent waves when we zoom in. The symbols are different temperature-sensing wires – with thicker wires responding slightly slower, as expected, confirming the waves to be a real atmospheric temperature effect and not some electrical artefact on the probe. *(Author)*

and terminating in a 'blob'. This was much more suggestive of a surface-controlled structure.

The profile of the upper atmosphere made by recording the drag deceleration during the entry phase of the probe's descent showed some prominent wiggles. These seem likely to be 'gravity waves'. They are also seen at higher altitudes in Cassini INMS data, and are even hinted at in data from stellar occultations. After the parachute came out, the probe measured pressure and temperature directly, and a close look at the temperature data showed beautiful wave activity above ~60km.

The intention had been for the Doppler profile of the probe's transmission during its descent to be measured continuously, several times per second, by Cassini. Unfortunately, a commanding error meant that the ultrastable oscillator that was to measure the received frequency accurately wasn't powered up. As a result, the measurement was lost. The problem was realised when the downlinked 'data' was found to be blank! But several large radio telescopes were prepared to observe the transmission. They could (just) make the measurement, but their record was much sparser and had large gaps. Nevertheless, the

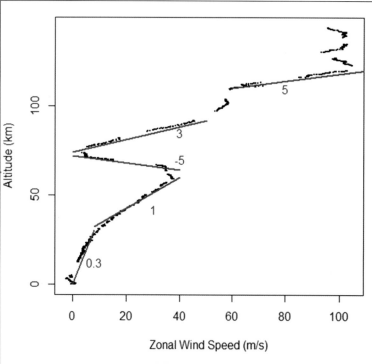

LEFT The zonal wind speed profile measured by Doppler tracking (black dots). The many small gaps were caused by the radio telescopes (first at Green Bank in the USA, then Parkes in Australia) 'nodding' to a calibration target. The bigger gap around 10km occurred when Titan was too low in the sky for either telescope to track it. The blue numerals indicate the shear in metres per second per kilometre indicated by the red lines. *(Author)*

key result was achieved. The winds were in the prograde direction, as expected, and at a speed also broadly about what was expected from the Voyager temperature data. Lower in the atmosphere, the zonal winds were a little weaker than expected, and in fact for most of the lowest 7km the flow was weakly retrograde. A significant finding was a shear layer in the stratosphere where the wind dropped to almost zero.

The gradient in the wind profile (that is, the wind shear) was in fact rather modest. Even the steepest parts, with a gradient of 5m/s/km, would in terrestrial aviation only be labelled 'light' turbulence (the shear would need to reach 17.5m/s/km to qualify as 'moderate').

Huygens also measured winds by correlating features seen in multiple images, permitting the position of the probe to be back-calculated. This gave the only good information on the north–south winds which were, as expected, small (<1m/s). Interestingly, this direction was reversed in the lowest few hundred metres, being southward near the surface: this effect, reproduced in circulation models, is most likely due to the season, in which the surface flow was toward the southern (summer) hemisphere, although it is impossible to rule out some local effects due to topography.

A challenge to modellers has been to get the zonal circulation right. Doing this properly may require fully three-dimensional simulations, which are expensive in computing time. Some models got almost everything right apart from the zonal superrotation. The Huygens profile gave modellers a firm target to aim for, and infrared observations by Cassini throughout the mission showed (albeit at a lower vertical resolution) how the winds varied with latitude and season. Just as the jetstream is stronger on Earth in winter, so Titan has a strong stratospheric jet where the zonal wind intensifies at the edge of the winter polar atmosphere. As with Earth's ozone hole, this jet helps to 'bottle up' the polar winter chemistry that increases the abundance of nitriles on Titan (see later).

The infrared spectra from the CIRS instrument enabled the variations of temperatures in the stratosphere (as well as at the surface) to be calculated in terms of latitude and altitude, and assuming that the corresponding latitude gradient in pressure is balanced by a gradient in centrifugal force of the rotating winds, these data allowed the zonal winds to be mapped at several times during the mission (as against just a couple of points by Voyager). Strictly, the method failed over the equator, but the largest and hence most interesting variations were in the high latitude jets anyway.

Methane and hydrogen profiles

The methane profile on Titan echoes the water vapour profile on Earth. Near the surface the abundance is probably rather variable, depending on whether it has rained recently or if the local surface is a sea or a desert, etc. On Earth, the concentration of water vapour is typically about 1%, which corresponds to around 50% relative humidity at 20°C. The Huygens surface value of 5–6% is ~50% relative humidity for methane at 94K. The concentration declines in the upper troposphere because the lower temperatures cannot support high abundances without forming clouds and removing the vapour.

Since water vapour on Earth and methane

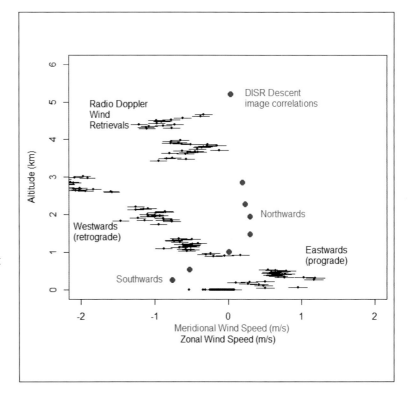

BELOW The near-surface profile of zonal (E–W) winds measured by Doppler tracking of the Huygens probe (black dots with horizontal error bars) and meridional (N–S) winds from its trajectory, as reconstructed using camera imaging. *(Author)*

on Titan are destroyed by solar ultraviolet light at high altitudes – making ozone and haze respectively – their abundance in the stratosphere is determined by how much leaks up from the troposphere to replace that which is destroyed. This in turn depends on the coldest temperature encountered at the tropopause, serving as a 'cold trap' throttling the upward flow. On Earth (today) the ~217K temperatures at 11–20km altitude clamp the stratospheric water vapour at a tiny amount – just a few parts per million (ppm). On Titan, however, the sphincter is less tight and it allows ~1.5% of methane into the stratosphere.

This fundamental difference is a big part of the reason Earth is more than half covered in deep oceans, while Titan has only dregs of shallow seas. On Titan, methane can escape quickly and be lost, whereas the water on Earth is sequestered on the surface and in the lower atmosphere, almost hidden from solar ultraviolet. But this distinction will evolve because as our Sun turns into a red giant, further warming Earth, the cold trap will weaken, allowing water vapour into the stratosphere where it will be quickly lost. This process appears to have happened to Venus early in solar system history. Hence, Titan today is a window on the processes that determine the fate of oceans.

The third most abundant gas in Titan's atmosphere is hydrogen, at 0.2%. This small molecule with a relative molecular mass (RMM) of 2 is much more easily able to escape Titan's gravity than can methane (RMM=16) or nitrogen (RMM=28), and so its presence requires continuous production. This is (probably) the same process that destroys methane: photochemistry in the upper atmosphere. As methane is converted into heavier hydrocarbons and nitriles, there is a bit of hydrogen left over, and this readily escapes continuously to space, making the methane destruction irreversible.

The hydrogen abundance had been estimated from Voyager spectra, and was measured. But there was a hint in the Huygens measurements that the hydrogen abundance might reduce a little toward the ground. That would happen only if there were some process on the ground which was consuming hydrogen. Chemical reactions run very slowly at 94K, but in principle the hydrogenation of acetylene, which is likely to be abundant on the surface, could yield free chemical energy. If an exotic form of life on Titan were metabolising hydrogen and acetylene, one signature might be a depletion of hydrogen close to the ground. However, the observed signature is tiny (and possibly subject to instrumental

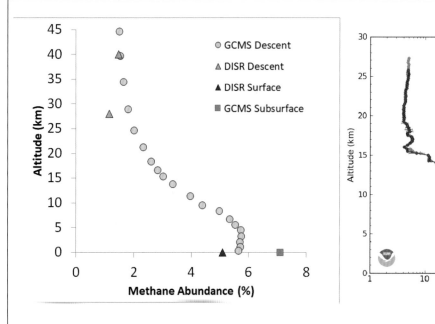
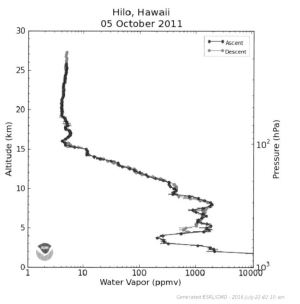

BELOW The methane profile on Titan roughly follows the saturation profile through the troposphere, except for a well-mixed region near the ground. The near-constant value in the stratosphere is set by the 'cold trap' temperature at an altitude of ~40km. The shape resembles the profile of water vapour on Earth (right), but methane is more abundant, and the altitude scale is larger on Titan. *(Author/NOAA)*

effects) and hardly the 'extraordinary evidence' that one would require for the extraordinary claim of 'weird life' on Titan. Even so, the Huygens observation will serve to motivate measurements on future missions.

Haze

A key objective for Huygens was to measure the optical properties of the haze from inside the atmosphere at different levels: the retrieval of surface and atmospheric properties from Cassini and ground-based observations depended on these details, which had only been estimated from laboratory analogues and assumptions about how the haze worked.

Inevitably, some assumptions proved to be incorrect. Most prominent was the expectation that condensing organic gases would coat the haze particles at an altitude around 20–80km, causing them to grow and hence fall faster, or 'rain out'. This process would make the lower part of the atmosphere relatively clear of haze.

That appears not to happen (or at least, had not happened in the recent past at low latitudes). Although Titan's surface loomed into view as the probe descended, this occurred slowly as the probe looked through less and less atmosphere, not because the atmosphere contained any less haze. In fact, Titan's atmosphere isn't actually very hazy at all – it only looks that way because when viewing from outside we are peering through 150km of the stuff. Meteorologists define 'visibility' as the range at which the contrast falls from 100% to 2%. A terrestrial 'fog' is a visibility of less than 1km and a 'haze' is in the range 2–5km. Huygens showed that the lower atmosphere on Titan has a visibility of tens of kilometres in red and near-infrared light. Images taken on the surface (which look through almost no haze at all) had a contrast of about 12% – the diffuse lighting and perhaps a coating of some dust muting the local scene.

A more subtle point about the haze inferred from the carefully planned measurements of the solar aureole (how light was scattered around the Sun) at different altitudes, is that while the haze particles seemed as expected to be clusters of much smaller particles, the clusters were in fact much larger than the pre-

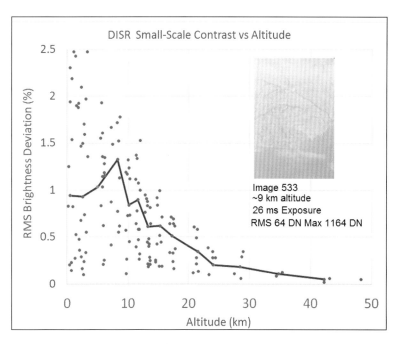

ABOVE The original Huygens DISR image data have rather low contrast (inset) and so are usually 'stretched'. As Huygens got closer to the ground the image contrasts (points, average shown by red line) increased significantly, although the last few kilometres of descent were over a blander area.
(Author from data by E. Karkoschka)

BELOW This part of the final VIMS colour ratio map is zoomed in around the Huygens landing site, marked by a red cross. The Cassini data (principally from flyby T47) matches up nicely with the Huygens DISR mosaic from the parachute descent, with the bright triangular 'highland' region well defined. The main image is about 800km across. *(S. Le Mouelic/LPGNantes/VIMS)*

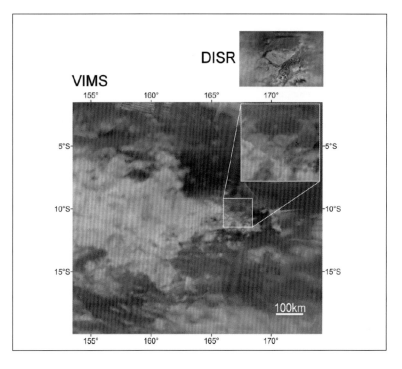

Cassini models anticipated. This may in part have been because the properties of these 'fractal aggregate' particles are much harder to calculate than the small clusters. The result was that the haze is a little less opaque at shorter wavelengths and has a larger opacity toward longer wavelengths. There was a hint that the particles became a little brighter at all wavelengths descending from 200km to 80km, and then that below 55km each particle intercepted a bit more light (i.e. it was optically 'bigger'). Hence, although wholesale rainout did not appear to be happening, condensation processes did modify the particles as they descended through the stratosphere.

All things considered, however, the pre-Cassini work did an excellent job of estimating the environment that Huygens would encounter in terms of winds, light levels, and so on.

Uncovering Titan

The data from Cassini came slowly and unevenly. There were several definitive revelations from the first couple of flybys, such as the surprising complexity of the organics in the upper atmosphere. Then the Huygens data took centre stage. But after that, Cassini's observations largely had to follow the plans laid out in the years of work during the cruise to Saturn, with only limited opportunities for refinements in response to initial findings. However, over time data progressively mounted up, and the opportunities to optimise in response to discoveries could be exploited.

The architecture of Cassini's tour around the Saturnian system drove how the groundtracks were distributed over Titan. Because Cassini used Titan's gravity to redirect its path around Saturn, such as making the orbit larger or smaller, or cranking the inclination of its orbit, the different stages of the tour in turn drove the character of the Titan flybys. Thus, after the first few Titan flybys wrested Cassini's orbit down into the ring plane for studies of several other moons, from T8 to T14 the Titan groundtracks were equatorial. Then to drive the inclination back up, Cassini flew over the polar regions from T16 onward, and so on.

The original design of the Cassini mission was for four years, making 44 flybys of Titan (plus T0) through summer of 2008. But scientists hoped that Cassini's fuel might last longer, and indeed the flight controllers worked

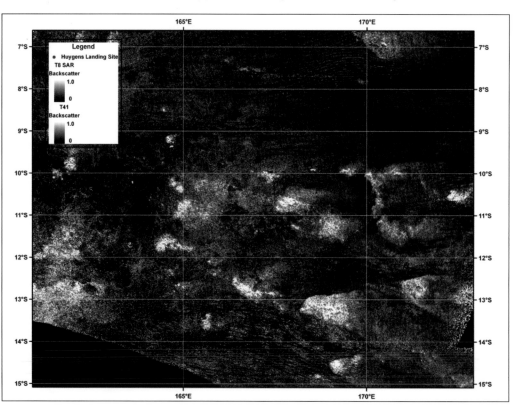

RIGHT The Huygens landing site (red dot) was imaged by Cassini's radar on T8, about eight months after the descent, and again on T41. These views gave a wider context for the probe's mosaic. Although the probe's immediate environs were a streambed, broadly speaking Huygens landed in a dune belt peppered with mountains (inselbergs). The image is about 350km across.
(M. Malaska)

hard to husband this and other resources to permit an extended mission. Special measures were taken to minimise wear and tear on systems like the thruster valves and the reaction wheels (big flywheels) used to orient the spacecraft. One ageing effect that planners could do nothing about was the slow decay of the plutonium-238 heat source of the radioisotope thermoelectric generators (RTG). But the slow decay in output, driven in part by the 78-year half-life of this isotope, was predictable – eventually the number of instruments operating during a flyby would have to be reduced, but this was a modest limitation.

Random failures occurred of course, as in any complex engineering system. The Cassini Plasma Spectrometer developed a short circuit in the early 2010s, and for safety was turned off. Another plasma instrument lost some scanning ability when a motor failed. And after its own internal ultrastable oscillator failed in 2011 the Radio Science Subsystem was reliant on reference signals from Earth. For the most part, however, these issues caused only modest reductions in capability and occurred well after the design life of the relevant systems (which had operated in space for some 11 years by the end of the original 'nominal' mission).

As things turned out, the Cassini mission was extended twice. First, a modest two-year extension was introduced with a 'business as usual' pace of activities spanning the equinox in 2009. This 'Equinox Mission' gave an additional 26 flybys of Titan and featured a preponderance of flybys over its southern hemisphere, which had been rather less well observed than the north during the nominal or 'Prime' mission.

Then, as Cassini's scientific productivity continued and scientists' appetites grew, the daring plan to extend operations through the northern summer solstice in 2017 was developed. This, however, would require a more relaxed pace, with less frequent flybys so that a reduced staff could manage operations (and thus minimise costs). Similarly, great care would be needed to eke out both the fuel for orbit changes and the propellant used for the thrusters that made fast turns. Whereas in the prime mission during a mapping observation the thrusters might rapidly ping-pong the spacecraft back and forth to cover as much terrain as possible in the tight flyby timeline, now scientists would accept rather less coverage by slewing the spacecraft using the reaction wheels. It was essential to husband propellant to reduce the likelihood of running the tanks dry before the planned end of mission.

Although the first radar image, which covered ~0.5% of Titan's surface, demonstrated that the instrument could resolve fine details, it was rather inscrutable in that few of the landforms seen were unambiguously interpretable. It was known that at the large scale observable from Earth, the microwave and optical reflectivity of the surface were correlated, Xanadu appearing bright in both. This correlation seemed to hold for some smaller-scale landforms, but by no means all: evidently Titan's surface was complicated. The optical and near-infrared albedo is effectively a measure of how dark the surface is 'painted', while radar interrogates the surface to a depth of some tens of centimetres and indicates its roughness.

Further information on the surface was afforded by VIMS data. Although a full near-

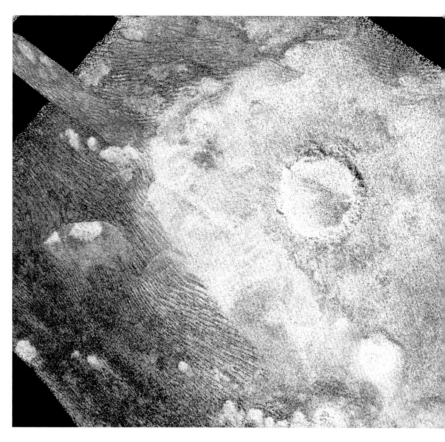

ABOVE **The 80km diameter impact crater Sinlap, the freshest-looking crater on Titan, was imaged by radar on the T3 flyby, just a month after the Huygens descent. Dark dunes sweep around the crater ejecta blanket, and the foreshortening of the crater wall in the oblique radar view indicates that the crater is 1,600m deep, making it Titan's deepest.** *(NASA/JPL/ESA)*

infrared spectrum gave rich information on the composition of Saturn's moons and rings, at Titan much of the spectrum was blocked out by the methane absorptions of the atmosphere. Instead of the 256 wavelengths on the VIMS detector, only a few dozen contained information on the surface of Titan.

The windows at 1.6, 2 and 5μm gave snippets of spectral information (beyond the albedo, or average brightness) from which to attempt to infer the surface composition. For example, at 2μm, a narrow absorption band of ethane that was sufficiently distinct from similar methane absorption, implied that Ontario Lacus, the only major lake in the southern hemisphere, had significant amounts of ethane. Similarly, efforts were made to attribute the shape of the 5μm absorption seen in the 'brown' unit associated with the dunes of Titan to particular organic compounds. However, the rich array of chemical possibilities and the limited spectral resolution of the instrument precluded a specific identification.

In principle the relative albedos at different wavelengths also constrain composition, but the effects of the haze in the atmosphere must be taken into account because it both scatters and absorbs light to differing degrees according to wavelength. And these effects also vary with both the angle from vertical at which the observation is made ('emission angle') and the angle between the view direction and that to the Sun ('phase angle'). A number of approaches have tried to correct these effects, ranging from ad-hoc to sophisticated, but disagreements among spectroscopists about the effectiveness of different corrections have made it difficult to draw significant conclusions.

As the principal information is present in three wavelength bands, it is natural to portray the albedo data in false-colour maps with the three primary colours representing different near-infrared wavelengths. In the most well-established scheme, a prominent unit at low latitudes shown as brown indicates an organic-rich composition associated with optically-dark regions where dunes have been observed. Another unit shown in blue has been interpreted to imply a composition relatively rich in water ice. Although it had been expected that perhaps Xanadu (and other bright material) would be

BELOW A selection of orthographic views from the global VIMS map colour ratios. The upper-left image is centred at 0°, 170°E, close to the Huygens landing site. The equatorial dunefields appear readily in brown. Blueish material indicating water ice is exposed at a few impact craters and other locations.
(S. Le Moueilic/VIMS)

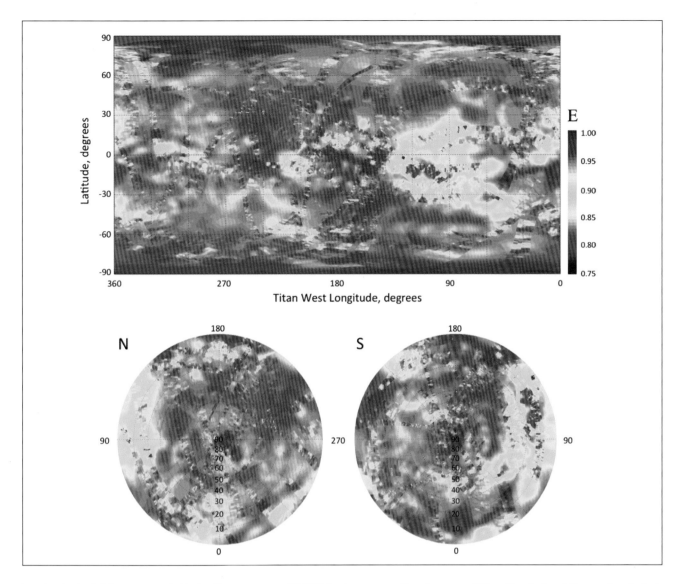

ABOVE **A microwave radiometry map of Titan. The dark red (i.e. most emissive) areas are the dunefields, plains, and seas. Xanadu and an area to the southeast are striking anomalies, indicating a distinct composition and/or texture.**
(M. Janssen/A. LeGall/NASA/JPL)

ice-dominated, this appears not to be the case. The blue unit is associated, it seems, with areas where rivers might have recently deposited sediments carved out from ice-rich bedrock which is itself covered up with something else. These appear in particular to be found on the eastern margins of bright areas – a relationship which has not been conclusively explained but presumably is in some way related to wind (perhaps, 'brown' sand migrating eastward readily covers up the western margins of the brighter, possibly more rugged terrain). The blue unit also appears prominently in the ejecta blankets of a couple of relatively fresh impact craters where (just like the bright rays of craters on our Moon) deeper material excavated from beneath a darker veneer has not been exposed for long enough to be modified.

MICROWAVE EMISSIVITY

Materials that have a higher dielectric constant (typically, with a higher density) will have a weaker thermal emission, expressed in terms of a 'brightness temperature' or 'emissivity'. A perfect emitter ('black body') at Titan's physical temperature of 94K would have a brightness temperature of 94K and an emissivity of 1. A deep ocean of liquid methane with a dielectric constant of 1.7 would have an emissivity of 0.99. Solid organics would have an emissivity of ~0.95. Water ice with a dielectric constant of ~3.1 would have an emissivity of ~0.92.

 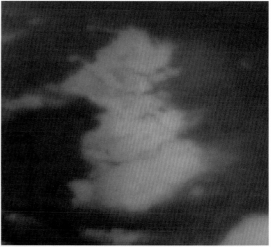

RIGHT Shikoku Facula was an ISS-bright region (right) west of Xanadu, in the Shangri-La dunefields (dunes are visible in the radar image at the left). While some features like dark wandering channels match up in the two datasets, others (like the 'eye') do not, showing that these instruments are telling us different things about the surface. *(Author)*

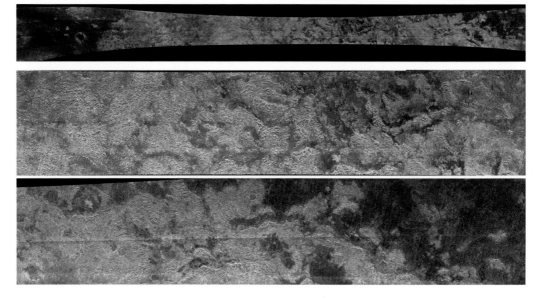

RIGHT The T13 radar swath (top) extended ~5,000km from the Shangri-La dunefields (containing the circular structure Guabonito) eastward across Xanadu. Zoomed sections show rugged mountains and many river networks. In the lower panel are several circles that may be degraded impact structures. *(Author)*

The suggestion that recent ejecta blankets contain material rich in water ice is supported by the radar on Cassini operating as a radiometer. The microwave emission is somewhat less affected by surface texture than is the reflectivity and hence can constrain the composition of the upper decimetres to decametres of the surface.

The first radiometry observations on the TA encounter revealed that much of Titan's surface is covered in organics. As higher-resolution data accumulated, a few regions emerged with low emissivities compatible with icy surfaces, in particular the ejecta blankets of the impact craters Selk and Sinlap.

The emissivity of Xanadu, the first distinct terrain to be spotted on Titan by astronomers, was very low (less than physically plausible for solid ice), suggesting a peculiar textural as well as compositional character. It is an area of paradoxes. The fact that it is both optically bright and radar-bright prompted the pre-Cassini suspicion of an ice-rich area; perhaps a highland swept clean of dark organics by enhanced mountain rainfall. The only part of this speculation borne out seems to be that it is mountainous, since radar images show rugged blocks of mountains. Yet it is not a highland – the locally steep mountains seem to be superposed on lower terrain. Furthermore, the microwave radar/radiometry properties introduced an extra wrinkle in that the polarisation of radiation is diagnostic of composition and says whether radiation has been reflected once at a highly reflective surface (like a mirror or white paint), or has been scattered multiple times by inhomogeneities inside the volume of material (like the glitter from a frosty lawn, or a milky ice sheet). The characteristics of the polarisation of the radar reflectivity and microwave emission are such that Xanadu and the area to the southeast must provide a lot of volume scattering.

Xanadu appears to host many of Titan's observable impact craters. Several of them are very degraded in character – although their broadly circular imprint is obvious, there is little vestige of either a rim or ejecta blanket. One possibility is that these structures were buried long ago, then disinterred. Perhaps if Xanadu had been covered by a lot of material then, just as ice sheets once did over northern Europe, that would have had the effect of loading the crust sufficient to form an overall depression. Clearly there is a complicated story in these terrains which remains to be unravelled.

Some features located to the south of Xanadu attracted study throughout the Cassini mission. Attention was drawn to the first, Hotei Regio, in early VIMS observations that showed it as a ~400km-wide spot at 80°W, 20°S with a distinctive spectrum that was particularly bright in the 5μm window. Indeed, upon review, it was evident in images from the Keck telescope in 2003. As active 'cryovolcanism' on Titan with 'lava' temperatures of 273K (for water without salts or other antifreeze) or even 176K (water with the maximum amount of freezing point depression by ammonia) would yield appreciable thermal emission at this longest VIMS wavelength, the enticing speculation was that it was a volcanic 'hotspot'. However, the rather low microwave brightness temperature seemed to rule out the prospect of present-day activity.

Curiously, although the spot itself is not prominent at 940nm, the southern edge of the feature shows up in the ISS map as a bright arc. This prompted speculation that a circular ridge was somehow controlling the presence of a deposit (perhaps old lava) inside. Radar images of the area were eagerly sought, and when finally obtained in 2008 they showed an arc of hills, with little river channels draining toward the centre. Blobby textures inside the spot seemed to be consistent with some kind of flows. Hotei is now believed to be a lakebed.

Questions of nomenclature deserve some discussion at this point. Firstly, what is the

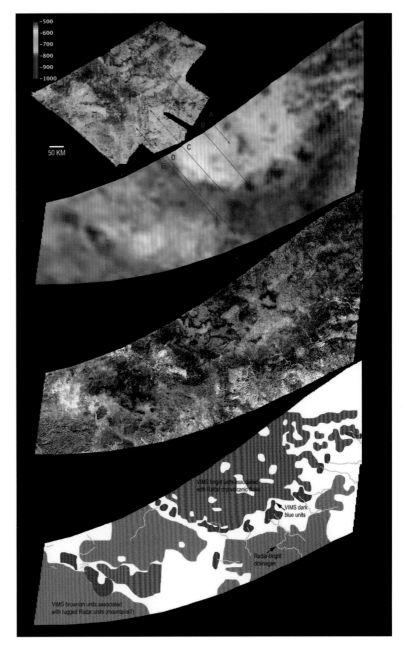

RIGHT **False-colour VIMS data (upper-middle, with an interpretive map at the bottom) indicates the variety of materials in Hotei Regio. The high resolution of the SAR image (lower-middle) shows lobate structures, and river channels draining radially toward the centre of the arc. Topography from stereo radar (top) helped with the interpretation. Hotei is now believed to be a lakebed.** (NASA/JPL/U. Arizona)

definition of lava? Obviously, it is stuff that is heated in the interior, comes out of the ground and flows. But that definition could apply to hot springs on Earth. So one has to exclude material that has participated in the hydrological cycle of being evaporated and then rained out. But terrestrial geology gives us some situations that defy even this seemingly straightforward classification. In Iran, for example, there are features where material coming out of the ground forms glacier-like flow features, and indeed the flow rates have been measured at about a metre per year. But the material is not molten, it is a rather soft rock (in this case salt) that flows. Massive layers of salt were deposited hundreds of millions of years ago when this area was an ancient sea (ironically, given the Saturnian theme here, known as the 'Sea of Tethys') that dried up, and in fact the salt layers are instrumental in trapping the petroleum deposits that occur in that part of the world. As

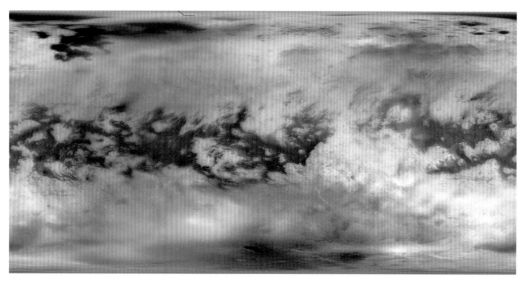
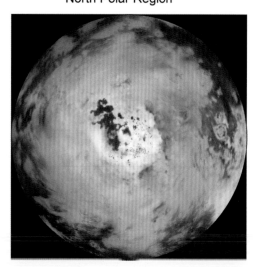
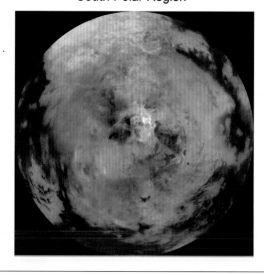

BELOW A global ISS mosaic of Titan. The south and north polar regions had to be made using images in different seasons (at the beginning of the mission, the north was in darkness). The cylindrical map was built up from all available data with techniques to compensate for different lighting conditions. *(USGS/SSI/E. Karkoschka)*

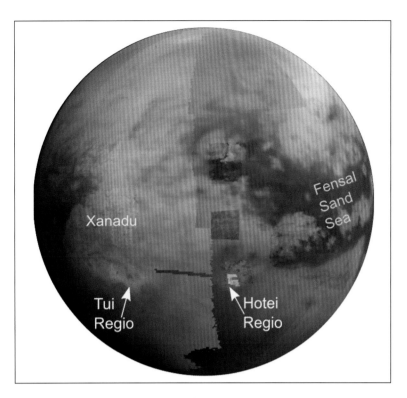

RIGHT **A mosaic of VIMS data, projected onto a sphere, with the red image channel corresponding to 5μm brightness. Two very prominent areas are Tui Regio and Hotei Regio.** *(S. MacKenzie)*

sediments have piled up on top of the soft salt, it has been squeezed up through cracks and pipes to 'fountain' out onto the surface in what, to us, is super-slow motion but is very rapid in geological terms. The material is not lava, and yet a planetary geologist might – to judge from the geomorphology alone – not unreasonably suspect the feature to be a volcanic flow. Given the plethora of possible organic compounds on Titan, it would be easy to imagine material that would be solid but could still flow when buried and subjected to a small amount of heating. In fact science fiction anticipated the idea – in his novel *Imperial Earth*, Arthur C. Clarke describes the eruption of a 'waxworm', forming what is effectively a lava tube.

So we must entertain a range of possibilities. As the Cassini mission went on, and more high-resolution VIMS coverage built up, it discovered many smaller areas of 5μm brightness. These were often at the margins of lakes (including Ontario Lacus), or in depressions that might have once been lakes or seas. Thus they resembled 'bathtub rings' and salt flats in terrestrial deserts, the bright deposits of once-dissolved minerals which were exposed as the liquid level dropped by evaporation.

It is entirely possible that the 5μm bright material at Hotei is different from that which forms the evaporite deposits (and Occam's razor has

DEGREES OF CONFUSION: LONGITUDE

A surprisingly vexing matter in planetary mapping is that of longitude convention. While there is typically little argument about the placement of zero longitude (in contrast to on Earth where Greenwich and Paris vied for the honour), the *meaning* of the number can be critical. To explain, geographers and geologists (people who like maps) tend to say positive is westward. However, geophysicists working on topography, gravity and magnetic fields tend to prefer the 'right hand rule' of the physical sciences (i.e. with the thumb northward, positive follows the fingers) which defined positive as eastward, a format that is a little more convenient for plotting software. One should always check!

In the case of Titan, zero longitude is defined as the sub-Saturn point at periapsis at a certain epoch. A further issue concerns the map centring. In part because there was a bit of a gap in coverage, the original Hubble map of Titan ran 360 to 0, west longitude, with the anti-Saturn point in the centre and Xanadu on the right. However, many VIMS maps used a 'Greenwich' convention of zero (sub-Saturn point) in the centre, with Xanadu on the left. The fact that the Huygens landing site and the Selk crater lie off to the edge in such a projection, means it will probably be used less and less.

Some attention is required when discussing landforms affected by meteorology: an 'easterly wind' is one that blows from the east, so its direction is 'westward'!

Latitude is generally uncontroversial, although in precision work it must be specified whether geocentric or geodetic latitudes are used (the former assumes a spherical body, the latter takes the ellipsoidal shape into account).

Furthermore, precision mapping work such as tying particular reflections in Arecibo data to Cassini maps must state the spin model that was used when generating the coordinates. The discovery of Titan's 0.3° obliquity and polar precession can 'displace' features by up to about 10km.

RIGHT **Views from space of salt glaciers in Iran. Were Cassini to observe Earth, would we imagine that this material was laid down on a seabed, buried and finally squeezed back out of the ground? Titan could be just as complicated. The feature in the bottom image spans about 14km, and occasional rainfall has cut wiggly channels into the soluble material.** *(NASA)*

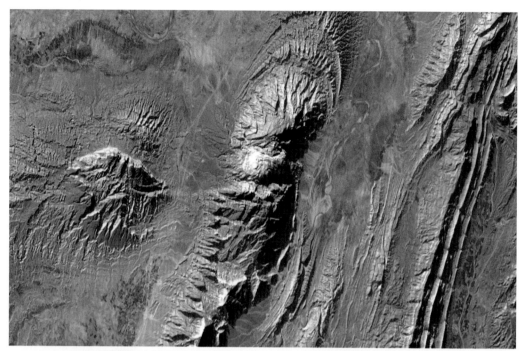

proven to be an unreliable guide in developing our understanding of Titan), but the consensus from the data at hand is that Hotei is some sort of lake deposit. Quite how the lobe textures formed, whether by erosion, or by 'waxworm'-like flows, is not known. Another aspect of the lakes is also still puzzling – although many in the northern 'lakeland' region sit at the base of 200m steep-walled irregular depressions, these pits possess elevated rims. It appears almost as if something 'erupted' from them. Perhaps the rims are the equivalent of the 'sinter' that is found at hydrothermal systems such as the geysers at Yellowstone National Park or around the mud volcanoes of Azerbaijan.

This chapter has sought to express some of the bafflement of planetary scientists, confronted with the slow dribble of data about a world that was bewildering in its complexity, and with exotic circumstances and materials. But as data accumulated over the years, an at least partly coherent picture began to emerge.

RIGHT One way to enhance interpretation is to overlay the low-resolution false-colour composition data from VIMS onto the higher-resolution morphology from radar. The fact that the floors of many dry lakebeds in the north and Tui Regio are bright at 5μm hints at an evaporite deposit of some kind. *(J. Barnes/VIMS)*

BELOW A mosaic of the Synthetic Aperture Radar (SAR) imaging from Cassini, covering about 45% of the surface at resolutions of better than ~1km. The large mid-latitude gaps result from the orbital tour of Cassini around the Saturn system, which precluded ground tracks from these locations. *(USGS/A. Hayes)*

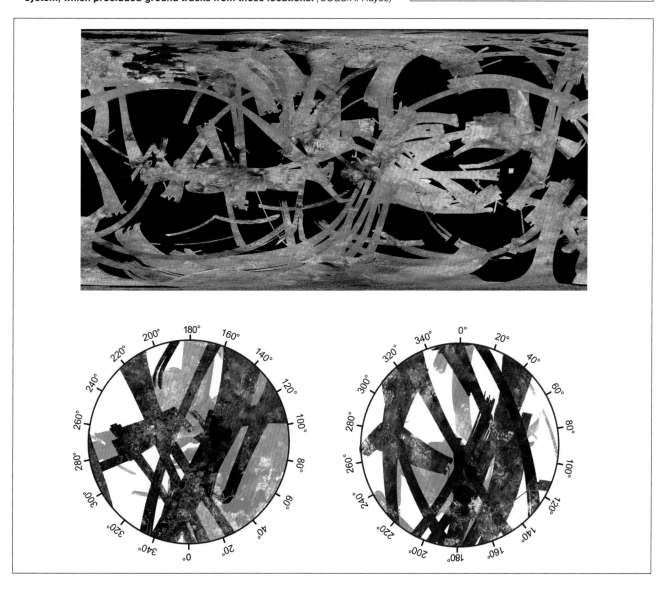

TERRAIN TYPES

Six main terrain types were defined by a study in 2019 that drew together assessments from all of the Cassini maps.

Terrain	Coverage	Description
Plains	65%	Areas lacking obvious topographic expression or other features. Lack of river valleys suggest porous ground. Generally ISS-dark and VIMS 'brown' suggesting organic sand but colour/brightness is variable.
Dunes	17%	Vast fields of dunes 1–2km wide, 2–4km apart. High emissivity suggesting organic composition. VIMS 'brown', ISS dark.
Hummocky	14%	Rugged and often radar-bright materials, mountain chains. Xanadu is the 'type locality'. Often VIMS 'blue-ish'.
Lakes	1.5%	Includes seas, and dry lakebeds.
Labyrinth	1.5%	Locally elevated and highly incised terrain.
Crater	0.4%	23 craters >20km diameter identified with certainty, plus ten 'probable'. Rim area only (e.g. floor may be plains/dunes etc.).

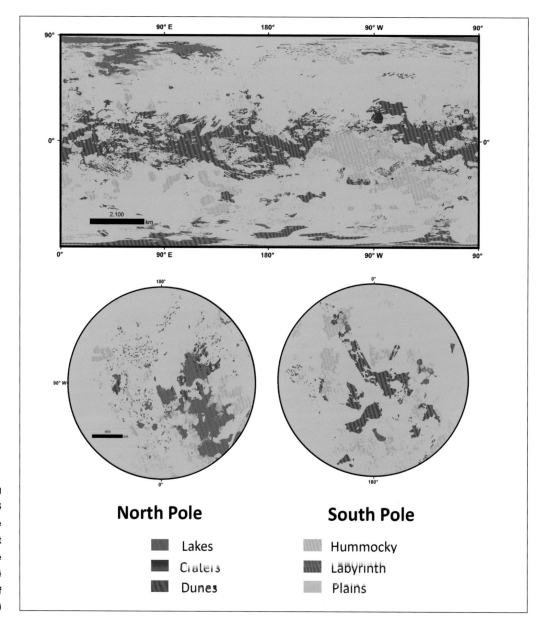

RIGHT By combining radar, VIMS and ISS data, geologists have classified different terrain on Titan. The result reveals the complex diversity of this world. *(R. Lopes)*

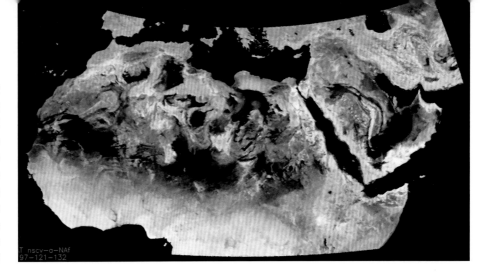

RIGHT The radar reflectivity of the Earth's surface is affected profoundly by vegetation and moisture, but the large-scale distribution of sand and mountains is evident in this mapped satellite scatterometry data from North Africa and Arabia. The darkest areas are flat lakebeds that reflect the radar energy away and sand seas that absorb it. *(BYU/MERS/D. Long)*

NAMING TITAN FEATURES

Scientists must inevitably call features something as they discuss them soon after discovery, so informal designations (e.g. 'Circus Maximus' for Menrva, the 'Heart of Darkness' for Belet, 'H' for Fensal-Atzlan) are used for a while. But 'official' names for landscapes on other worlds are eventually set by the Committee on Planetary Nomenclature of the International Astronomical Union. It ensures that a wide range of cultures are represented in the sourcing of names, with proposed names being appropriate, reasonably pronounceable, not easily confused with other planetary features, and inoffensive.

Prior to Cassini's arrival, Titan's features (even the well-observed bright region that became Xanadu) were not assigned names because their nature was not at all understood – a prudent policy, in retrospect! The first 43 names were approved in 2006 and more than 200 more have since been allocated.

Table: Albedo features

The following name themes were adopted for the different feature categories on Titan. So far, the lakes theme has not led to confusion in the field of limnology, but this could happen one day!

Terrae (Lands)	Sacred or enchanted places, paradise, or celestial realms from legends, myths, stories, and poems of cultures from around the world
Colles (Hills)	Names of characters from Middle-Earth, the fictional setting in fantasy novels by English author J.R.R. Tolkien (1892–1973)
Craters and ringed features	Gods and goddesses of wisdom
Facula and faculae (Bright spots)	Facula: Names of islands on Earth that are not politically independent / Faculae: Names of archipelagos
Fluctūs (Flows)	Gods and goddesses of beauty
Flumina (Rivers)	Names of mythical or imaginary rivers
Freta (Straits)	Names of characters from the Foundation series of science fiction novels by American author Isaac Asimov (1920–1992)
Insulae (Islands)	Names of islands from legends and myths
Lacūs and lacunae (Lakes and lakebeds)	Lakes on Earth, preferably with a shape similar to the lacus or lacuna on Titan
Maria (Seas)	Sea creatures from myth and literature
Montes (Mountains)	Names of mountains and peaks from Middle-Earth, the fictional setting in fantasy novels by English author J.R.R. Tolkien (1892–1973)
Other features (maculae, regiones, paterae, and arcūs)	Deities of happiness, peace, and harmony from world cultures
Planitiae and labyrinthi (plains and labyrinths)	Names of planets from the Dune series of science fiction novels by American author Frank Herbert (1920–1986)
Sinūs (bays and inlets)	Names of terrestrial bays, coves, fjords or other inlets
Undae (dunes)	Gods and goddesses of wind
Virgae (streaks)	Gods and goddesses of rain

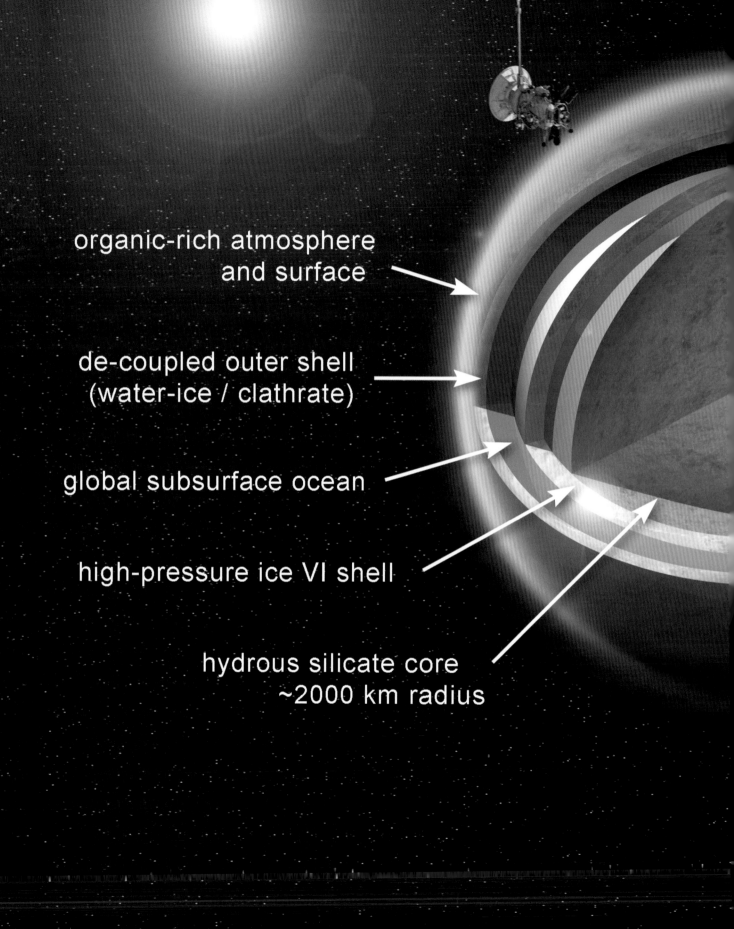

Chapter Four

Evolution and interior

Titan's interior may host a vast ocean of liquid water beneath an ice crust. Even a hundred miles thick and supporting mountains a mile high, this crust deforms measurably. Buried beneath a veneer of organic material, the ice is occasionally excavated by impact craters.

OPPOSITE A model of Titan's interior based on Cassini gravity measurements. Beneath the orange organic-rich surface and atmosphere is an icy crust (grey) overlying a global water ocean (blue) that may or may not be isolated from the rocky core (yellow) by a layer of ice of a kind that forms only at high pressures (white). *(NASA/A. D. Fortes/UCL/STFC)*

ABOVE **The moons and planets were assembled from a disc-like cloud of rock and more volatile materials. Farther from the hot centre, volatiles like ice were more abundant.** *(NASA)*

The worlds of the solar system are composed of varying amounts of iron, rock, and ice. They were assembled by a process of collision in a cloud of such materials that surrounded the Sun ~4.5 billion years ago. This 'protosolar nebula' had gradients of temperature and composition, with the cooler parts farther from the Sun being richer in water ice, especially after the 'snow line' located a little beyond the orbit of Mars. Thus while Mercury, close to the Sun, is a dense 'iron planet', the outer planets, made from more volatile materials, have much lower densities.

The giant planet systems formed in their own little 'protoplanetary nebulae', which had their own gradients. As a result, Jupiter's four large moons have quite different compositions, with the innermost, Io, being a dense (3.55kg/m^3) body of iron, rock and sulphur that is volcanically active. Next outward, Europa, has a veneer of about 100km of water and ice which reduces its density to ~3kg/m^3. The outermost of these moons, Ganymede and Callisto, have a density of about 1.9kg/m^3 that suggests they are about half rock and half water – the latter mostly in the form of ice. These moons are known as the Galileans because their existence was reported by Galileo early in the 17th century.

With a density of 1.88kg/m^3 Titan was expected to have a broadly similar composition, but its atmosphere betrays a higher abundance than the Galileans of some more volatile materials like ammonia, methane and nitrogen – as might be expected from the fact that the proto-Saturnian nebula was cooler than that around Jupiter. It was these conditions that made Titan so special by giving it an atmosphere.

The major mystery prior to the Cassini mission was where did Titan's atmosphere come from, and why is it still there? A key part of the puzzle, indicated instantly in the first Huygens and Cassini results, was the abundance of the different isotopes of argon. Being a noble gas, this is not processed chemically. It is therefore a good tracer of the physical processes of atmospheric evolution. Argon comes in several atomic masses that have different origins: Ar^{36} is produced by nuclear fusion in stars, and almost all the argon in the Sun and in Jupiter is of this form. In contrast, Ar^{40} is released by the radioactive decay of potassium (specifically K^{40}) in rocks, and the argon that forms 1% and 2% of Earth's and Mars' atmospheres respectively is almost all of this type. Argon and molecular nitrogen have similar volatility, and so if Titan's nitrogen had been delivered in that form (perhaps trapped in a water ice structure called a 'clathrate') then argon ought to be rather abundant as Ar^{36}. But the early Cassini INMS data and the Huygens GCMS showed only traces of it. However, they did find rather higher abundances (40ppm) of Ar^{40}.

This meant that the nitrogen on Titan must have been delivered in a less volatile form, almost certainly as ammonia. Titan likely had an early thick, warm atmosphere of ammonia and some processes (perhaps lightning and/or shockwaves from large impacts) converted the ammonia into nitrogen. Thus, in terms of getting an atmosphere, Titan 'cheated the system' – conditions were too warm at its formation to permit getting nitrogen, and conditions now are too cold for an ammonia atmosphere, but Titan converted its atmosphere from one to the other. It appears likely that the environment in which this conversion occurred would also drive a lot of organic synthesis that converted methane into heavier compounds and/or native carbon – which might explain some other Titan mysteries.

The isotopes of carbon and nitrogen add to this story, although here the interpretations are less clear-cut because these abundances can be influenced slightly by chemistry. In fact, the ratio of $^{14}N/^{15}N$ was measured by millimetre-wave spectroscopy using radio telescopes several years in advance of Cassini's arrival. The ratio in molecular nitrogen measured by Cassini turned out to be a factor of ~2 higher, presumably because some of the chemical pathways in the formation of hydrogen cyanide (HCN) have a slight isotopic preference. In any case, the $^{14}N/^{15}N$ ratio on Titan of about 200 is between that of Earth and that of Mars. It is believed that much of Mars' nitrogen was lost to space. As the lighter isotope would more readily escape, the greater the loss, the more the gas that remained would become enriched in the heavier isotope. Similar arguments involving the isotopes of hydrogen in water vapour suggest that Venus once had oceans, but over time they escaped to space.

Titan's formation history is, of course, somewhat speculative. Also speculative is

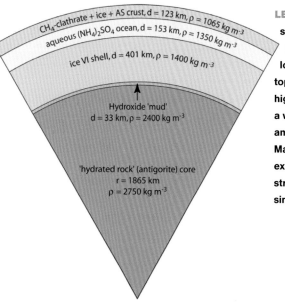

LEFT A possible structure of Titan, with a rather large, low-density core topped by a layer of high-pressure ice and a water ocean rich in ammonium sulphate. Many other models exist, but their broad structure is generally similar. *(A. D. Fortes)*

exactly how the 'half rock, half water' mixture is distributed today. Titan is big enough that as the ice-rock component pieces were drawn in by the accumulating object's own gravity, this occurred with sufficient energy to melt any ice. The rock parts likely had a composition similar to meteorites, and being denser they fell to the middle. The energy released in their arrival impact, and from their descent, is slowly leaking away, but it has been supplemented by the heat released by the radioactive decay of potassium, uranium and thorium in the rock. There might also be a bit of tidal kneading. The stresses caused by Saturn's gravitational pull, which varies around Titan's slightly elliptical orbit, dissipate energy in Titan. This is not nearly as much as for Io or Europa in the Jovian system though. The present-day heat flow leaking from Titan is guessed at about only 5mW/m². Although this is less than 10% of the average for Earth, it may well be sufficient to maintain a thick layer of water

BELOW Ocean Worlds – satellites with best-guess layers of different ices (light blue), rock (brown) and iron (grey). The liquid water oceans of Ganymede and Titan are 10–20 times the volume of the Earth's oceans (or that of Europa), which are in turn about 200 times bigger than that of tiny Enceladus. *(Steve Vance/JPL)*

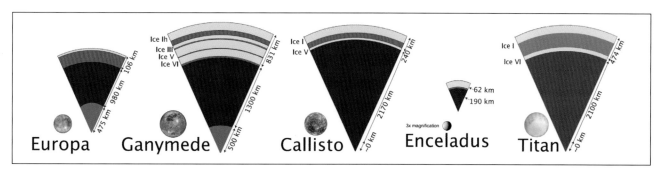

beneath Titan's icy crust to the present day. But whether this ocean is salty, or ammonia-rich, nobody knows. And how much of the ice crust is 'pure' water ice versus clathrate hydrate (a type of ice in which the crystal lattice acts as a cage to trap other molecules such as methane or ethane) we also do not understand. Perhaps the most important unknown is the thickness of that ice layer.

For the Galilean moons, some of the best information on this question came from the way that the changing magnetic field of Jupiter induced electric currents in an ocean. That in turn gave the moons their own weak magnetic fields. The Galileo spacecraft was able to detect these on close flybys. But this technique will not work for Titan because the magnetic field of Saturn is almost perfectly symmetrical at the radius of the moon's orbit and so does not vary. Also, the moon's thick atmosphere prevented Cassini from getting close enough to make the necessary measurements, and in any case there were confounding currents in the upper ionised parts of the atmosphere.

The first indications were from a rather obscure measurement by the Huygens probe, which measured a background 'hum' in its electric field sensors. This may have been a Schumann resonance, a kind of electromagnetic ringing occurring in the cavity between the conductive ionosphere and the conductive water ocean. Although the detection may have been simply mechanical noise on the probe, if taken at face value it hinted that perhaps the ice crust was only about 50km thick.

The next piece of evidence emerged from measurements of Titan's spin. Just as the rotational dynamics of a hard-boiled egg differ from those of a raw one, so planetary spin dynamics can provide insight into its interior. It had been presumed that Titan rotated with its equator in its orbital plane, which lies within 0.3° of the Saturnian equator and indeed the ring plane. It was expected that Titan rotated synchronously, always presenting exactly the same face to Saturn. The pre-Cassini telescopic observations seemed essentially consistent with this picture. These assumptions formed the basis for calculating the latitude and longitude of features seen with the radar instrument, which accurately measured features relative to the spacecraft and its motion. But as the coverage grew with each flyby, by late 2007 there were many overlapping parts of the long, thin radar image swaths and it was noticed that sometimes the features did not quite match up – a mountaintop seen during one flyby at one position would appear to be displaced by as many as 10km when observed again a few years later. Clearly, the mountain had not moved in relation to the rest of Titan. The problem had to lie in the assumed frame of reference.

Radar scientists then performed a feature-matching exercise in which the rotation state of Titan was a set of free parameters and the job was to calculate what position of Titan's pole and what rotation rate would give the best match of positions. The solution was somewhat degenerate, in that one couldn't tell with the data at hand whether Titan was rotating slightly asynchronously (one

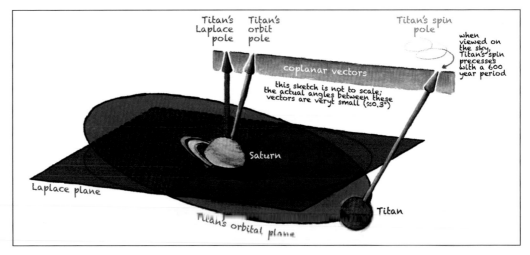

RIGHT Titan's Cassini State. The spin axis of Titan and the vector normal to its orbital plane are both coplanar with the invariant or Laplace pole, which is essentially the rotational pole of Saturn. *(James Tuttle Keane)*

thousandth of a degree per day faster than the 22.577°/day orbital synchrony), or the pole was precessing in the sky. Nevertheless, it was established that the pole was not quite where it had been presumed to be – instead Titan's rotational pole was offset by about 0.3° from its orbit normal. (This is very small compared with the 26.7° obliquity of Saturn and Titan with respect to the Sun, so the extra 0.3° has no meaningful effect on climate.)

This position was consistent with what astrodynamicists coincidentally call 'Cassini State 1' (after the astronomer). Over a period of ~600 years, the pole will describe a tiny circle in the sky. In fact, the pole likely moves in a cycloidal path. This behaviour (and indeed a possible nonsynchronous rotation that had been an early interpretation of the radar data) was much more likely if Titan were not a solid sphere of rock and ice but, as thermal evolution models predicted, an ice crust that was mechanically decoupled from the deep interior by a layer of liquid water.

A more established way of probing the interior of a planetary body is to use gravity. Several of Cassini's precious Titan flybys were devoted to gravity measurements, whereby the high-gain antenna was pointed at Earth for a strong signal and the use of thrusters which would disturb the measurement was minimised. These conditions prevented Cassini from making many other observations on those flybys. Careful analysis of the Doppler shift, measuring velocity changes of a fraction of a millimetre per second, yielded an indication of how much Titan's gravity field deviated from the perfect symmetry expected if the body were a uniform sphere. In particular, the field's shape as an ellipsoid is described by two numbers (called J2 and C22) in a harmonic series. These numbers were an indication of the distribution of mass inside the moon. What is more, they would change if the shape were to change. A clever experiment conceived during Cassini's development was to do the gravity measurement twice: once close to periapsis, then again near apoapsis – when the orbital eccentricity would increase Titan's distance from the planet by 3%. If the moon were absolutely rigid, its gravity field would be constant. But if it behaved more like

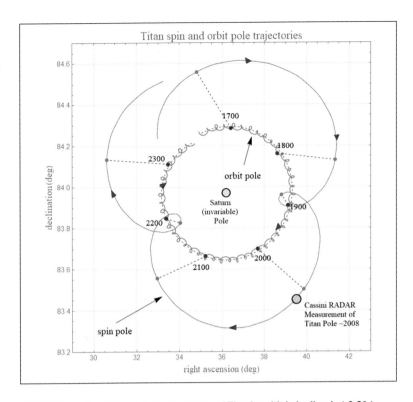

ABOVE Wheels within wheels. The plane of Titan's orbit is inclined at 0.3° to Saturn's equator. The gravity field of the oblate planet, and the effect of the Sun's gravity, cause Titan's orbit to precess in the inner spiralling circle over about 600 years (dates are for the snapshots indicated by the red dots/dashed lines). Titan's rotational pole, measured by Cassini between 2004 and 2008, is itself inclined at 0.3° to Titan's orbit pole, and wanders in a wide looping arc around the orbit circle. *(Bruce Bills)*

BELOW A schematic of Titan's tidal deformation. When it is nearest to Saturn (periapsis, at the upper-right), the stronger gravity gradient draws Titan into a more elongate ellipsoidal shape, whereas at apoapsis (bottom-left), the tidal gradient is about 10% weaker and the moon can relax into a more spherical shape (the difference is probably only about 10m, out of Titan's 2,575km radius and therefore has been greatly exaggerated here). *(NASA/JPL)*

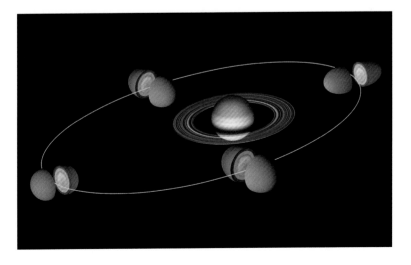

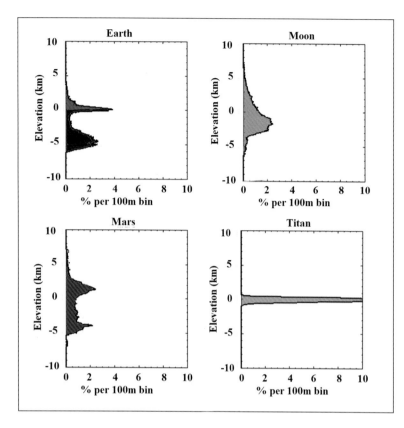

ABOVE Hypsograms show the relative amounts of terrain at different elevations. Earth has a bimodal distribution: more than half the planet's surface is ocean floor (blue), several kilometres below sea level, near where most of the land (green) lies, with only a tiny fraction of high land thrown up in mountain ranges. Without erosion to destroy the mountainous rims of craters, the Moon's hypsogram tails off only slowly to high altitudes. Mars is a hybrid between these two. The fact that Titan's histogram spans less than 2km probably indicates vigorous erosion, and possibly a weak crust that is unable to support large mountains. *(Author)*

a liquid (and an ice shell only a few per cent thick overlying a water layer would allow it to act this way) then the gravity field would differ by a measureable amount.

Several complications meant that a program of gravity flybys had to be conducted, so the final result indicating that the field changed by ~4% was not available until 2012. The interpretation is encapsulated in the Love Number k2 (named after a mathematician) which would be 0 for a perfectly rigid body and 1.5 for a perfectly fluid one. The value of ~0.6 firmly showed that Titan possesses a global internal ocean.

The deformation measured by Cassini was very small. For all practical purposes the crust of Titan is rigid. But just as the Earth's crust (or more correctly, its lithosphere) deforms slowly (Scotland and Scandinavia, for example, are still rebounding upward after the removal of the weight of ice sheets 10,000 years ago), the landscape can be influenced by the strength of the planet underneath.

Cassini's radar measured elevation profiles along its lengthy swaths, and by 2011 enough of these had accumulated to give some robust overall impressions. Apart from the mountains (seen at small scales in the radar images) Titan was pretty flat overall. The total topographic range was ~1.5km; only 10% that for Earth (~10km above sea level for mountains and ~5km below sea level for the ocean floor).

This small dynamic range could mean that Titan generates no mountains (but even our

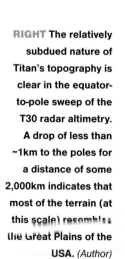

RIGHT The relatively subdued nature of Titan's topography is clear in the equator-to-pole sweep of the T30 radar altimetry. A drop of less than ~1km to the poles for a distance of some 2,000km indicates that most of the terrain (at this scale) resembles the Great Plains of the USA. *(Author)*

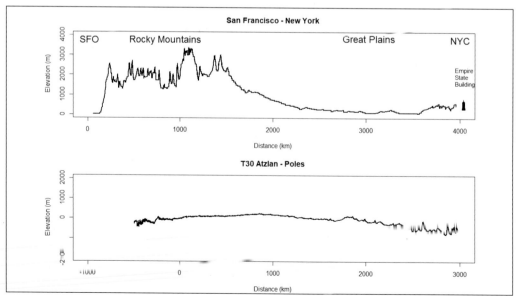

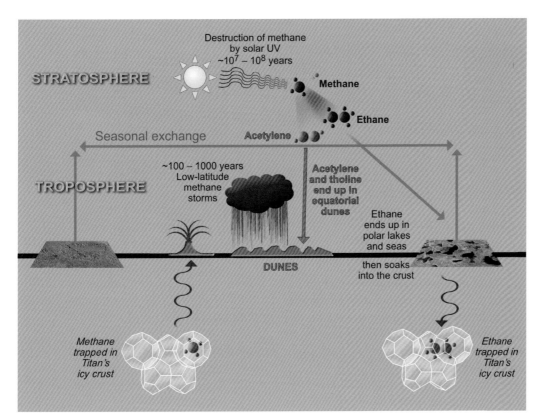

LEFT A schematic of the polar uptake of ethane into clathrate ice cages in the crust of Titan. Ethane would displace methane, partly replenishing the methane in the atmosphere that participates in the hydrological cycle. (NASA/JPL)

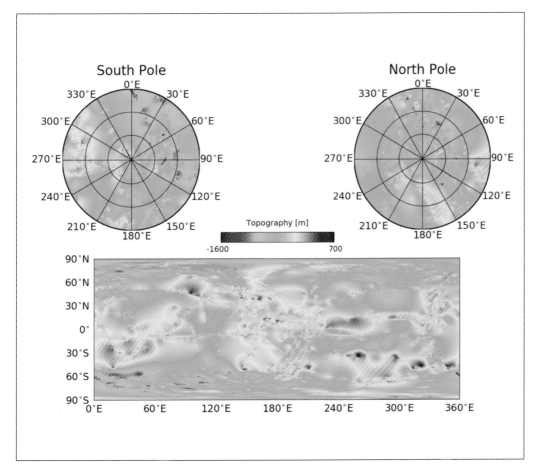

LEFT A topographic map of Titan from radar data. Poleward is generally downhill. Large smooth areas are gaps where the height is estimated by interpolating the surroundings. The deep depressions, including the lowest point known (lower-left), are possibly basins that once held seas. (P. Corlies)

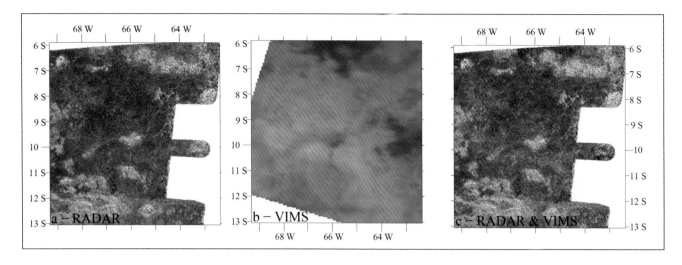

ABOVE Composite (right) image with VIMS colour data (centre) overlain on a radar image (left) of part of Xanadu. The mountains, and to some extent the thin bright river channels that radiate from them, seem to be associated with the blue colour, perhaps indicating that ice-rich material is eroding out of the mountains, which resist being covered with photochemical material. *(Jason Barnes)*

BELOW Titan's rift valley? The long straight edges of the two long sand seas near 45°W, Fensal (upper) and Atzlan (lower) suggest some kind of tectonic control, as if there were large-scale cracking, like that thought to form the massive Valles Marineris system on Mars. *(Author/J. Barnes/USGS)*

dead Moon has mountains that were formed billions of years ago by large impacts), or that erosion and burial by wind and rain grinds them down, or that the crust is not strong enough to hold them up. Probably it is a combination of the last two factors, but the topography observed on Titan suggests the ice crust must be at least 40km thick.

Both polar regions were found to be ~0.5km lower in elevation than the equator. This at least made sense, in that liquids should flow downhill!

But why should the poles be lower? One thought was that Titan has a 'frozen-in' shape from some time in the past when it was spinning much faster than it does today, now being tidally locked with its orbit. But there is little else to support this idea. An innovative suggestion was that in fact Titan's photochemistry and weather drives its shape. An uneven crust could be in gravitational equilibrium, floating on the underlying water ocean, if the crust were denser at high latitudes. Then the weight of the crust (its thickness multiplied by density) would still be the same everywhere. It is evident in images that some materials from the upper atmosphere preferentially condense at high

LEFT The T28 radar swath showed part of Ligeia Mare, plus the channel networks of Trevize Fretum and Vid Flumina. The fact that the centre-left shoreline of Ligeia is so straight, and that some of the channels are rather parallel and/or orthogonal, hints at possible influence of tectonic faults. The Moray Sinus estuary is to the left. *(Author)*

RIGHT The highest mountains known on Titan are the Mithrim Montes, seen here in a radar image processed with a technique to smooth out the speckle noise in the image. The mountain height has been estimated. *(NASA/JPL/A. Lucas)*

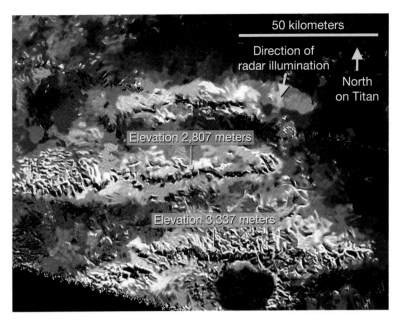

latitudes (forming the 'polar hood'). What if that material was making the crust denser at high latitudes? Perhaps ethane, a liquid expected to be abundantly present but so far only observed in small amounts, might percolate through pores in the ice to displace methane from a clathrate (experiments have shown that ethane *can* displace methane from such ice cages) making it denser.

Mountains and tectonics

One important question in the design of the Huygens probe was how much topography there might be on Titan. The specification was that the probe should have sufficient battery energy to support its operations for a minimum of 3min after the longest possible descent. Factors in the descent duration included the density of the atmosphere and the performance of the parachute, and also the height of the surface relative to the atmosphere profile measured by Voyager. The probe would descend at about 5m/s near the surface. If it landed on a 5km high mountain, the descent would be 1,000sec (~17min) shorter than nominal. On the other hand, if it were to land in a pit 5km deep it might exhaust its battery prior to reaching the surface. So how likely were surface elevations of +1km, etc.?

Mountains can be made in a number of ways, including the folding or buckling of the crust in response to horizontal stresses, and the thrusting up of blocks. Impact craters can also develop mountains – a central peak or

BELOW Mountains generally recognised in radar swaths, and assigned names from mountains in the Middle-Earth of JRR Tolkien's fantasy novels, are shown here on a VIMS global basemap. *(NASA/JPL-Caltech/U. Arizona/USGS)*

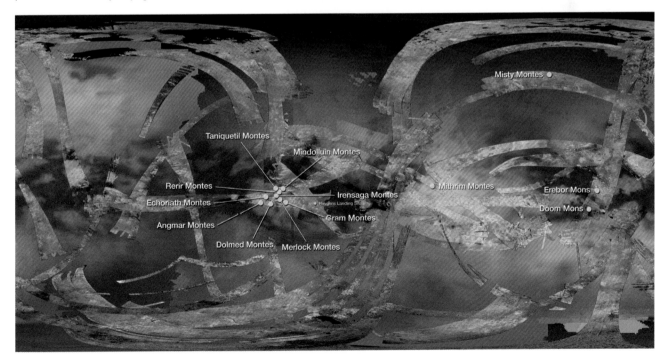

EVOLUTION AND INTERIOR

LEFT Mountain chains near Titan's equator, with dunes filling many of the plains in between, seen on the T8 swath. *(NASA/JPL)*

CENTRE The island of Mayda Insula, situated at the north of Kraken Mare, is similar in size to Jamaica or the Big Island of Hawaii. This topography model was developed from stereo radar observations, and shows a couple of mountains poking more than a kilometre above sea level, which is itself ~1,200m below the reference level on Titan. *(R. Kirk/USGS)*

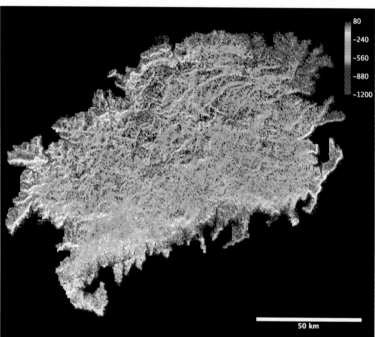

ring of hills, plus a raised rim. Most of our Moon's topography was made in this manner. Some of Titan's mountains (for example the parallel ridges of Mithrim Montes) seem to have been formed by the rotation of crustal blocks, rather like the basin-and-range region of the southwestern USA. Because many of the mountains on Titan are isolated radar-bright patches surrounded by sand dunes they are 'island mountains' – also known as 'inselbergs'. There are also some narrow mountain belts, but their formation mechanism isn't understood.

The most compelling candidate for a volcano (or rather, a cryovolcano because the rock is 'ice' and the lava is water, or at least water containing ammonia antifreeze) is Doom Mons. Radar topography revealed it to be rather steep, with a deep pit right next to it which is too small in diameter to be an impact crater. There is also something that looks like a flow emerging from the structure and running to the north. A good terrestrial analogue is the Galapagos Islands, where there are several examples of juxtaposed (volcanic) craters and mountains.

Craters

Impact craters have been found on every solar system body except Io, whose surface receives a volcanic makeover so frequently

LEFT Doom Mons is estimated from this radar stereo data to stand about 1,450m high with a diameter of about 70km. The (presumed volcanic) crater here at Sotra Patera is about 1,700m deep and 30km wide. The region also features finger-like flows, named Mohini Fluctus. *(NASA/JPL)*

RIGHT **A feature nicknamed 'The Snail' (left, now Tortola Facula) was observed by VIMS on the very first Titan flyby, and initially speculated to be of volcanic origin. A later radar view (right) found little to support this interpretation. The only reason VIMS happened to observe this spot was because it was the original Huygens landing site.** *(NASA/JPL/U. Arizona)*

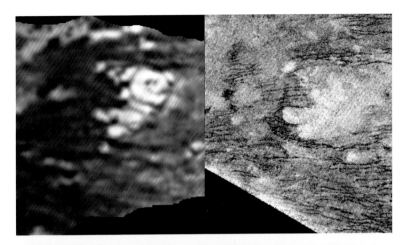

that such blemishes are blushed out. And there are craters on all of Saturn's moons apart from the portions of Enceladus that are mantled by fallout from its 'snow-gun' plumes, so it seemed natural that Titan would have many, and indeed that if there were surface liquids, they might form spectacular circular and ring-shaped crater lakes. But it transpired that Titan has relatively few craters, and in this respect its surface is much more like that of Venus or Earth.

One pre-Cassini prediction was right, at least. This was that there ought to be very few, if any, small craters. The reason is that the size of a crater scales with impact energy, and hence with the size (mass) of the impacting body, typically a comet or asteroid. But an object that is small enough to make a small crater on reaching the surface of Titan will be efficiently slowed by the atmosphere and may even break up. Thus, whereas the Moon has countless craters of all sizes, Titan has few smaller than about 10km in diameter – much like Venus, which also has a dense atmosphere that screens out small impactors.

Another result of Titan having a thick atmosphere is that ejecta from ground strikes cannot be thrown planetwide in thin rays, such as we see on the Moon. Instead, the cloud of debris from the impact explosion is trapped locally by the atmosphere and forms a thick blanket that may flow.

If the crater population on Titan were like that measured on the smaller neighbouring moons Dione or Iapetus by Voyager, then the Huygens probe might have seen one or two,

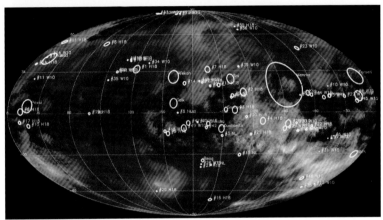

ABOVE **A map of Titan's craters. Circles are scaled to three times a crater's diameter. Most appear to be at low latitudes and are found preferentially on the leading (Xanadu) hemisphere (the right side of this map, in an equal-area Mollweide projection). Numbers denote as-yet unnamed craters in different catalogues.** *(Author)*

RIGHT **On Titan the density of surviving craters of a given size is 2–3 greater (about 1 per million km^2) than for Earth, and far lower than for Mars or the icy moons.** *(Author/C. Neish)*

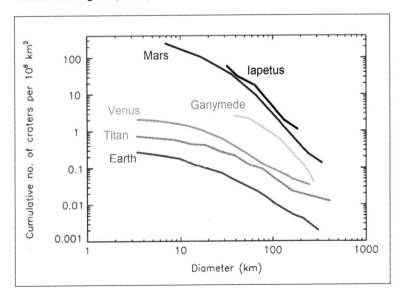

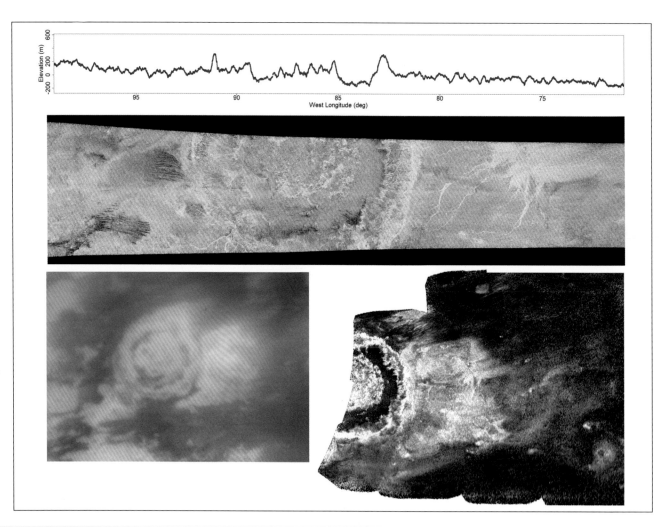

ABOVE Titan's largest crater Menrva, viewed in the T3 radar swath (centre) is 440km across, yet has very little vertical relief (radar topography profile, top). A mosaicked near-infrared view from ISS (bottom left) highlights the presence of dark sands – also visible in the radar image where it forms dark forked streaks, dunes. A second, low-resolution radar view at a very oblique viewing angle highlights surface roughness, and also caught the smaller crater Ksa (bottom right). *(Author)*

LEFT The snow-covered ice surface of a ring-shaped lake defines the eroded form of the 60km wide Manicouagan impact crater in Quebec, Canada. Before Cassini, it was thought that many such crater lakes might exist on Titan, but no good examples were found. *(Author)*

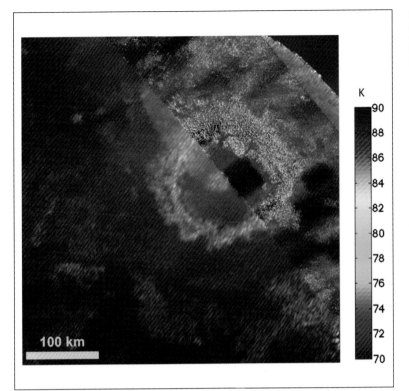

ABOVE **The Forseti impact crater. Imaged at relatively low resolution overall, it shows a degraded rim, a dark (sand-filled?) floor and a central peak complex. The upper-right part was imaged at higher resolution, showing the incised texture of the rim. The colour-coding displays the microwave brightness temperature or emissivity. The blue signature of the rim/ejecta suggests the excavated material is rich in water ice.** *(Alice Le Gall)*

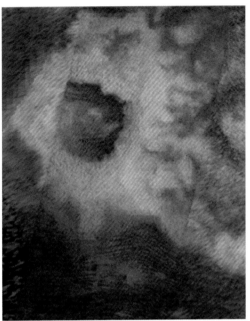

ABOVE **The ~70km impact crater Selk will be explored by the Dragonfly mission in the 2030s. It will initially land in the dunefields to the south. Radar data (higher resolution strip across the middle) is here coloured with VIMS data, the blueish tint indicating areas possibly rich in water ice. Like Meteor Crater in Arizona, Selk is appreciably square in shape, perhaps due to crustal faults that predated the impact. Some dunes can be observed in the crater floor.** *(Shannon MacKenzie)*

LEFT **The Roter Kamm impact crater in Namibia has a diameter about double that of Meteor Crater in Arizona, but, like Titan's craters, it is much shallower because it has been filled in by windblown sand and occasional rain action. This view was taken by a GoPro camera that was suspended from a parafoil kite.** *(Author)*

RIGHT The Roter Kamm crater interacts with Namib dunes, the sand (yellow-green in this near-infrared false colour composite) all but covering the crater. Alluvial gravels wash down from the hills in the upper right. *(NASA/METI)*

RIGHT Seen by radar, Roter Kamm resembles many craters on Titan, with just its bright rim being visible. The green points in this multi-wavelength image are succulent bushes which are bright because the length of the leaf stems happen to be very close to the radar wavelength. Pink colour in upper left is characteristic of reflection by fine gravel. *(NASA/JPL)*

and there would have been dozens in each Cassini radar swath. But the first radar swath (TA) yielded none – immediately suggesting that Titan might have a surface that was geologically young overall. (Actually, a short profile strip taken by the radar on TA in its non-imaging altimetry mode did show a rise which eventually turned out to be the rim of a crater, but no one knew that at the time.)

Some vaguely circular features were seen in the early Cassini ISS images, notably a ring that was 400km across. This was observed more closely on the second radar flyby (T3), the month after Huygens, and indeed seemed unmistakeable as a crater. Named Menrva, it remains the largest impact crater confidently identified on Titan.

One characteristic of Titan's craters is their shallowness. In comparison to craters of the same size on Ganymede, whose gravity is similar and presumably has broadly similar mechanical properties, the craters on Titan for which the depth has been measured are a factor of 2 or so shallower. Presumably rainfall and rivers degrade their rims, and together with wind-blown sand (some dunes are visible in the larger craters) help to reduce the depth of the cavity. The same T3 flyby that revealed Menrva also imaged a rather pristine crater, 80km across, named Sinlap. The radar imaging geometry allowed its depth to be estimated at 1.3km; it may be the deepest crater known.

A striking feature of the rims and ejecta of some craters on Titan is they have very distinctive remote sensing signatures. The structures correlate with a VIMS 'blue' spectral colour that is suggestive of water ice. They also look 'cold' to the Cassini radar receiver when it senses the black body microwave emission from the surface (a measurement that the instrument makes 'for free' between

BELOW A 'zoo' of craters imaged with Cassini's radar – hardly any of Titan's craters look alike. In many of the examples shown here, dark sand dunes intrude from the west, with white arrows indicating apparent edges of ejecta blankets. *(Author, adapted from work with Catherine Neish)*

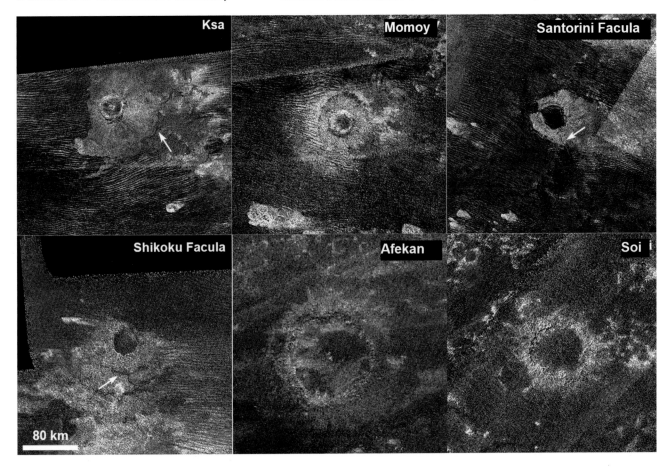

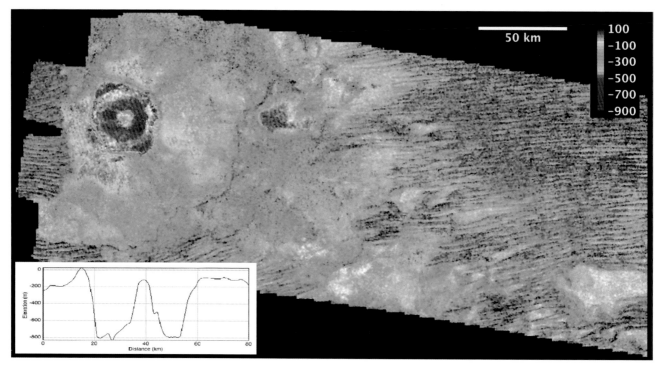

ABOVE A stereo radar model of the crater Ksa (colours show elevation in metres). The ~600m-tall central peak almost rises to the level of the crater rim. Note the dunes to the right. *(Randy Kirk, USGS)*

BELOW The craters on Titan are shallow in comparison with those of the same size on Ganymede or the Moon. *(Author, adapted from work with Catherine Neish)*

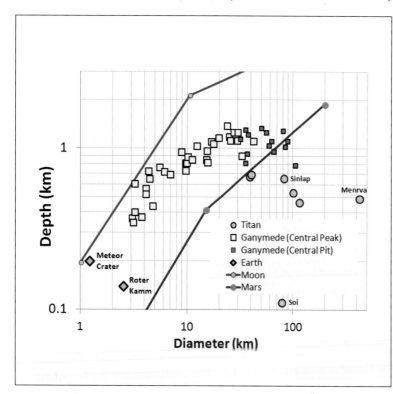

radar pulses). One material with the requisite electrical properties is water ice. This had been expected to be the principal crustal material, not silicate rock, but it seems generally to have been covered by some thickness of organic material.

Balance of power

The question of crater depth is a nice example of quantitative geomorphology. What makes a landscape? Having hidden its surface from Voyager's scrutiny, Titan preserved a mystery for scientists (this author among them) to test their understanding of planetary geomorphology. With the basic known properties such as gravity, atmospheric density, sunlight and material (assumed to be water ice), was it possible to predict the nature of the surface? My own forays into this arena were motivated by my work on developing instrumentation for Huygens – in particular, what might a suite of force sensors and accelerometers measure on impact? (This presumed that the probe would survive long enough to transmit such data, which was by no means guaranteed.) There were many plausible possibilities: would the probe land gently in fluffy organic dust; slam down onto an unforgiving sheet of solid ice; crunch into a bed of icy gravel; or splash into a sea of liquid methane?

It is fair to say these prognostications were not entirely successful. This is, however, the way science is supposed to work: a theory is challenged by observation, and that prompts a better theory. A major deficiency in the predictive efforts was that they treated Titan as if it were a single place, whereas in fact latitude proved to be a profound control on rainfall, and on the transport of sand. I have often been asked by TV broadcasters looking for a location to shoot Titan documentaries 'Where on Titan is like the Earth?' After Cassini's initial findings, I have taken to retorting 'Where on Earth is like the Earth?' There is no single defining landscape on our planet, except perhaps the open sea.

And so it has turned out on Titan. All the 'Titan Surfaces' that were speculated upon for quite logical scientific reasons before Cassini, proved to be present on Titan – but only to an extent. Although the possibilities were recognised, the diversity of features was not really anticipated.

One analysis that was published shortly prior to Cassini's arrival ignored what Titan might be made of, and just considered what processes shaped the surface. It evaluated 'work rates' for the atmosphere (aeolian transport and the hydrological cycle), endogenic processes (volcanism and tectonics driven by radiogenic heat), tidally driven processes, and impacts. For Earth, this calculation indicates a vigour of atmospheric-driven processes 100 times more than that of the thermodynamic work from tectonics, which in turn is three times greater than tidal effects, and 10,000 times greater than impact cratering. The observed area fractions of the types of features associated with the various mechanisms on Earth at least roughly matched their relative order of importance; that is, Earth's surface is a largely sedimentary one with a small area fraction of volcanoes and mountain belts and an even smaller fraction of impact craters.

On Titan, the processes were expected to be somewhat more balanced, but with atmospheric processes still dominant. And so it proved – Titan indeed had more craters than Earth, but a random location (the Huygens landing site) was evidently shaped by rainfall, and Cassini's surveys found widespread river valleys and sand seas.

Xanadu

The bright region on the leading face of Titan was recognised in lightcurve studies in the early 1990s, and in the 1994 Hubble map and subsequent observations. The naming authorities were initially reluctant to consider names for a feature whose origin and nature were not known but did assign the name 'Xanadu', redolent of mystery, to this region.

When it was delineated in 1994, it was speculated that perhaps Xanadu was elevated icy terrain that was washed clean of (assumed) dark organics by rainfall. However, while it was bright at a range of near-infrared wavelengths consistent with this picture, it was also bright at 5μm, which was puzzling because water ice is relatively dark at that wavelength.

It has been speculated that Xanadu is an impact structure. While it is a distinctive region to be sure, it is hardly circular, and there seems little geomorphologic evidence for an impact origin in the Cassini radar data. It is a

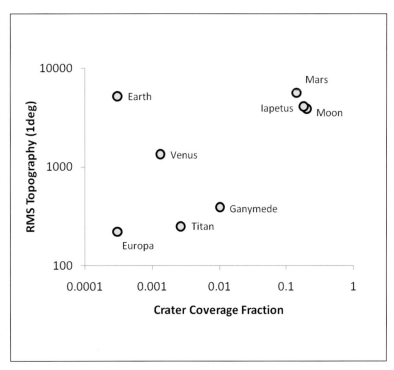

ABOVE A phase diagram of planetary geological activity. Worlds with few craters are either frequently, or at least recently, resurfaced (Earth and Europa, less so Venus and Titan), while geologically dead worlds retain the scars of their late formation (Mars and many moons of the solar system, including our own). Earth has a great deal of topography because it is still building it. The fact that Titan's ice crust is thicker than Europa's may partly explain how it is able to support more topography. *(Author)*

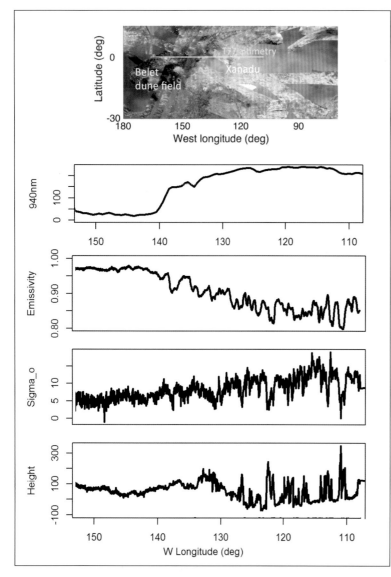

LEFT Shangri-La to Xanadu. Although Xanadu is bright to ISS (940nm) and radar, its boundary is less well defined in terms of radar reflectivity and microwave emissivity. Rather surprisingly, the topography data (bottom) indicates it has a low floor, with several hundred-metre-high mountains superposed on it. *(Author)*

characteristic of planetary science that impact cratering is the last resort of the scoundrel: if there is something that cannot be easily explained another way, one can simply wave one's hands and say 'a giant impact did it', and of course it is difficult to prove against such a *Deus Ex Machina* origin.

What is striking about Xanadu is that it is very rugged. SAR imaging shows it to be plastered with mountains. Some of the plains between these mountains are dissected by extensive river valleys – in fact Xanadu has the best-developed networks on Titan, showing several orders of branching. However, despite being mountainous, it is, overall, rather lower in elevation than the Shangri-La dunefield to its west. Which begs the question why the sand doesn't fill it up.

Xanadu also hosts many medium-sized impact structures, often in rather advanced states of degradation and yet rather crisply defined. Perhaps it is an ancient terrain that was buried in sediment for much of Titan's history and then disinterred.

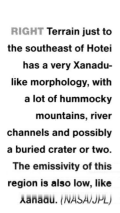

RIGHT Terrain just to the southeast of Hotei has a very Xanadu-like morphology, with a lot of hummocky mountains, river channels and possibly a buried crater or two. The emissivity of this region is also low, like Xanadu. *(NASA/JPL)*

> But indeed all the whole story of Comets and Planets, and Production of the World, is founded upon such poor and trifling grounds, that I have often wonder'd how an ingenious man could spend all that pains in making such fancies hang together. For my part, I shall be very well contented, and shall count that I have done a great matter, if I can but come to any knowledge of the nature of things, as they now are, never troubling my head about their beginning, or how they were made, knowing that to be out of the reach of human Knowledge, or even Conjecture.
>
> **Christiaan Huygens,**
> *The Celestial Worlds Discover'd*, 1698

The combination of microwave emissivity and radar data indicates that although this region is reflective, this reflectivity is not due to being a rough, solid ice surface – which would be more reflective than a smooth and/or organic one. Rather, there is some textural aspect of the terrain that makes it appear radar-bright; perhaps it is heavily fractured. As its fantastical name hints, Xanadu retains much mystery.

Titan's formation and orbit

It is conventional to assume that Titan formed in the way we think most moons formed, from a spinning cloud of dust and gas centred on the planet as that was itself forming. In this scheme, the angular momentum of the cloud forces all the moons to orbit in the same plane: any rogue moons in eccentric and/or inclined orbits will tend to collide with others and therefore not last long.

Broadly, Titan fits this picture, indicating it formed in the Saturnian system rather than being captured from interplanetary space (as Saturn's Phoebe, and Neptune's Triton, appear to have been). But a close look at the parameters leaves some puzzles.

Firstly, the architecture of the Saturnian system appears unusual (if comparison with the three other large satellite systems can be deemed 'usual'). Whereas the largest four satellites of both Jupiter and Uranus are of similar sizes in each system, Titan has no peer – in comparison to its closest rivals, Rhea and Iapetus, it has almost double the diameter and a mass an order of magnitude greater. This could be random chance, but it might suggest a different history.

Another puzzle is Titan's orbital eccentricity, which is an appreciable 0.029. While 3% doesn't sound like much (compared with, for example, Saturn's eccentricity around the Sun) it is much more than we would expect. Even if Titan started with a large eccentricity, energy dissipation in its icy interior, and especially in its surface seas (tidal dissipation in Earth's shallow seas is what makes the Moon slowly recede from us) could reasonably be expected to have damped out the eccentricity long ago. Either the eccentricity (and thus, perhaps, much of what defines the Titan that we see today) is relatively recent in origin, or some effect is maintaining it. It has been speculated that this might somehow be a result of orbital resonances with mighty Jupiter, but a more catastrophic mechanism is also a possibility.

Many geochemical and geological 'clocks' on Titan point to the relevant processes having been around 500 million years ago – only one-tenth of the age of the solar system. It would be a strange coincidence for Titan to have evolved in a uniformitarian way for 4.5 billion years, and for our understanding of all these different processes to be all wrong, and all by the same amount. The temptation that Occam's razor urges upon us is some sort of 'reset' half a billion years ago. But what? Perhaps an orbital rearrangement of the Saturnian system or some catastrophic collision.

Chapter Five

Rivers and dunes

Titan's landscape proved to be remarkably Earth-like, with dunes and rivers. Despite the very different working materials and physical conditions, the same physical processes which shape our planet's surface produce strikingly similar landforms on Titan.

OPPOSITE This aerial view over Namibia in southwest Africa shows parallel dune ridges sweeping past a mountainous obstacle that has spilled fans of alluvial sediment. Cassini observations show that Titan likely has identical vistas – minus the blue sky! *(Author)*

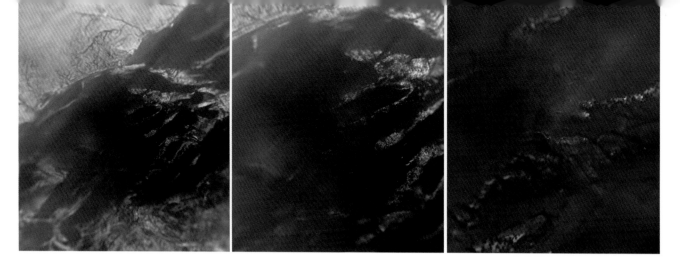

ABOVE Left: A Huygens descent mosaic from 10km altitude. The landing site is at the centre. Right: As the view from 1km shows, although the immediate vicinity of the Huygens landing site viewed from knee-height after touchdown seemed benign and flat, distressingly rugged gullies lurked a few hundred metres to the southwest. *(U. Arizona/ESA/NASA)*

On the foundations of an ice-rich crust that was forged on the large scale by impacts, tectonics and perhaps volcanism, the landscape on Titan, unlike airless bodies like the Moon or Europa, is worked by a dense atmosphere, rain, fluids, and the flow of methane rivers. These motions rearrange the ices and organics into landforms familiar to us here on Earth, especially in desert areas.

Rivers

Rivers are an example of the remarkable similarity in geomorphology between Titan and Earth. On one hand, Titan's low gravity makes it easier to lift a cobble or sand grain from the riverbed and transport it along; yet on the other hand the same low gravity gives less push for the liquid down a given slope. The net effect is that the ability to transport sediment for a given riverbed slope and depth of liquid ends up being rather similar. There are, of course, differences in the details, because ice and organics making up the sediments are less dense than terrestrial rocks, and liquid methane is less dense and less viscous than water. However, the basics remain the same.

The first close radar and optical images of Titan from Cassini in 2004 showed hints of narrow, bright or dark winding surface features, but it was the evidence from the Huygens probe that left no doubt. The images during the descent showed branching valley networks on the bright hills just to the north of the landing site and the post-landing images showed rounded cobbles littering a dark plain, implying a riverbed. Angular or tabular rocks are seen in almost every planetary landscape, but they can be attributed to impact cratering or volcanism. The presence of rounded rocks was an indication that they had been eroded as they tumbled in an energetic flow like a flash flood. Although measurements by the probe showed that the sandy material beneath the cobbles was damp with methane and ethane, there was no indication of pooled or flowing liquid. We cannot say how much time has passed since it last rained there.

Just weeks after the Huygens landing, Cassini's radar found spectacular examples of river channels near (and in) the giant crater Menrva. The network of filigree channels to the east showed repeated branching and recombination that is characteristic of shallow channels in desert environments on Earth. This occurs when the river flow is so energetic that it doesn't really care whether there is a pre-existing channel to follow, it just makes a new one, and it requires the heavy rainfall typical of the summer rainstorms in deserts. This particular river network was given the name Elivagar Flumina, after the ice-cold poisonous streams of Norse mythology.

The channels were mostly radar-bright, implying rough beds at the scale of the radar's 2.2cm wavelength, so they must be littered with pebbles or cobbles. But a few were dark, implying the presence of a smoother, more finely grained sediment, or perhaps a liquid. A nice example was seen in September 2008 when a densely incised network of canyons descended through one of the longest slopes measured directly on Titan (falling ~1.5km over

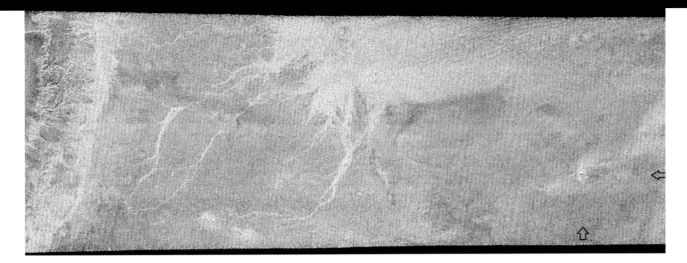

ABOVE **The braided network of shallow channels named Elivagar Flumina extends for hundreds of kilometres east of the Menrva crater, whose rim is on the left. The branching and reconnection of channels is characteristic of arid regions on Earth, where they are produced by sudden flash floods from rainstorms. The arrows point to a small feature once speculated to be a volcanic caldera.** *(Author)*

RIGHT **This extensive dendritic (branching) network of river channels in Xanadu flows to the south, with some individual channels traceable over 500km. Multiple stages of branching can be seen, just like river networks on Earth.** *(Author)*

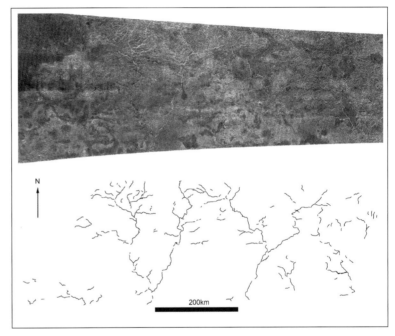

a distance of about 1,000km). In the lowest part the features branched out onto a plain and were dark, implying a distributary channel.

Several places on Titan seem to have alluvial fans. These are triangular fields of cobbles and boulders and they occur where a river which starts off confined and energetic in mountainous terrain is able to spread out across a shallow slope, dropping cobbles as it loses energy. Again, these features are characteristic of very episodic rainfall that causes the river flow to vary a lot (and typically stops altogether for long periods). Some of the best examples are seen in Death Valley in California but they are found in

RIGHT **A T7 view of a channel network draining southeast of Xanadu. In the equatorial highlands there is much dissection, with deep channels revealed by bright/dark pairing in the radar image as the rivers descend over a kilometre in elevation. Toward the south as the terrain flattens out the rivers become dark distributary traces, depositing fine-grained sediments.** *(Author)*

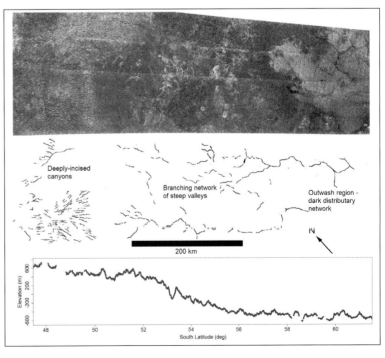

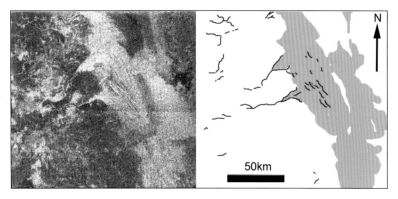
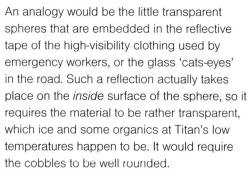

ABOVE **Leilah Fluctus on the TA swath, ~100km to the east of Ganesa Macula. The two triangular radar-bright channels and the associated bright area are interpreted as alluvial units where coarse, perhaps icy, sediments were deposited.** *(Author)*

RIGHT **A Space Shuttle radar image of Death Valley National Park in California showing alluvial fans spilling out from the mountains. The colours represent different wavelengths or polarisations of echo and respond differently to surface rocks of various sizes. The mountains appear tilted because of the radar ranging perspective. The little green dots and red streaks are sand dunes. At right are aerial views of alluvial fans, and (looking south) of the salt-flat valley floor.** *(Author photos; JPL radar image)*

many other mountainous environments, such as the Himalayas.

One example of channels emerging from a mountainous area on Titan drew attention because the riverbeds were exceptionally bright. As this was inexplicable in terms of mere roughness, a suggestion was retroreflection by radar-transparent spheroids.

An analogy would be the little transparent spheres that are embedded in the reflective tape of the high-visibility clothing used by emergency workers, or the glass 'cats-eyes' in the road. Such a reflection actually takes place on the *inside* surface of the sphere, so it requires the material to be rather transparent, which ice and some organics at Titan's low temperatures happen to be. It would require the cobbles to be well rounded.

It is striking that the channels extend some 100km from the mountains, so the cobbles must be able to travel that distance without being ground down into sand or dust. Yet the cobbles were rounded rather quickly in the mountains. This obliges us to think carefully about the different kinds of strength of geological materials. Perhaps on Titan they are malleable so that (like lead or wax on Earth) they are easily

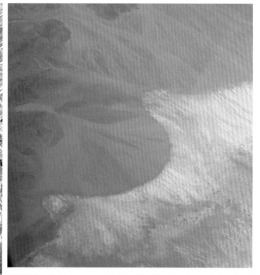

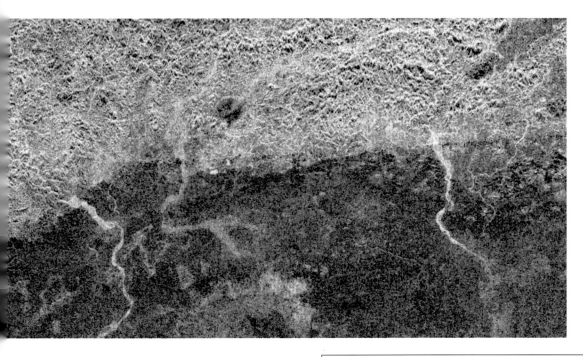

ABOVE The bright southwestern edge of mountainous Xanadu is evident in this radar image, but of note are the bright channels that extend ~100km beyond. Their startlingly high reflectivity points to very round cobbles several centimetres in diameter, implying that these particles can be transported by apparently meandering channels over a long distance without breaking down into sand. *(Author)*

BELOW Concentrations of remarkably well-sorted cobbles can result from violent fluvial transport, such as this 'jokullhap' (a catastrophic outflow prompted by volcanic melting beneath an ice sheet) on the south coast of Iceland. The Principal Investigator of the Dragonfly mission gives a sense of scale. *(Author)*

BELOW A schematic of the interaction of dune sands and river-borne sediments, which are coded as 'brown' and 'blue' respectively in VIMS maps. *(J. Brossier)*

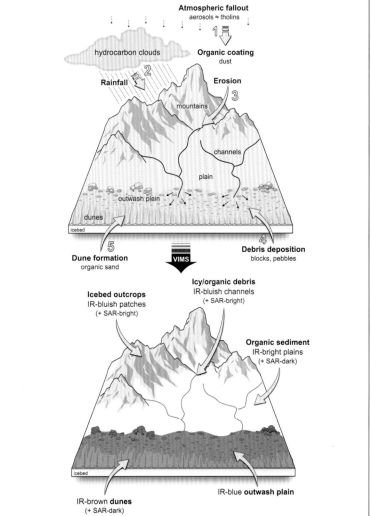

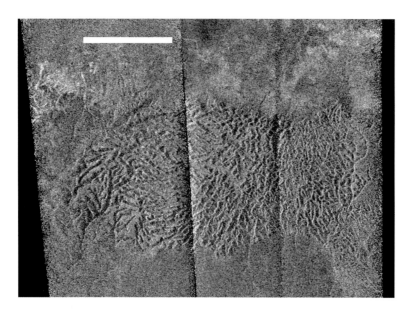

ABOVE The heavily dissected terrain of the Kaitain Labyrinth, not unlike a furrowed salt glacier on Earth. The ugly seam in this image (~200km wide) is due to incomplete correction for the strong elevation changes in the scene. Adjusting to correct for these seams yielded much of our topographic information on Titan. *(Author)*

BELOW A synthetic perspective of Sikun Labyrinthus, a chiselled plateau. *(M. Malaska/B. Johnsson)*

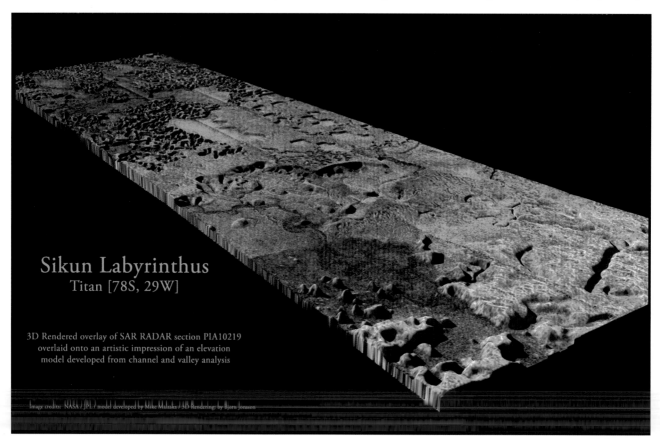

rounded without fracturing. Of course, these are relative terms because the energy associated with clashing rocks on the riverbed might be less in Titan's low gravity than is typical for Earth, making the material strength quantitatively rather lower than terrestrial rocks.

Canyonlands

Much pre-Cassini thinking about Titan was driven by first-principles physics and chemistry, and indeed, physical chemistry! With the assumption that the bedrock was water ice and the liquid was methane and/or ethane, one possibility was that the surface might dissolve to open sinkholes or caves similar to the Karst region of former Yugoslavia.

However, on Earth, river channels are carved into bedrock more by mechanical abrasion than by dissolution. Sand and gravel that is stirred by the flow will grind upon the bedrock surface. This type of mechanical erosion does not require the bedrock to be at all soluble in the flowing liquid.

A number of patches of Titan ('labyrinths')

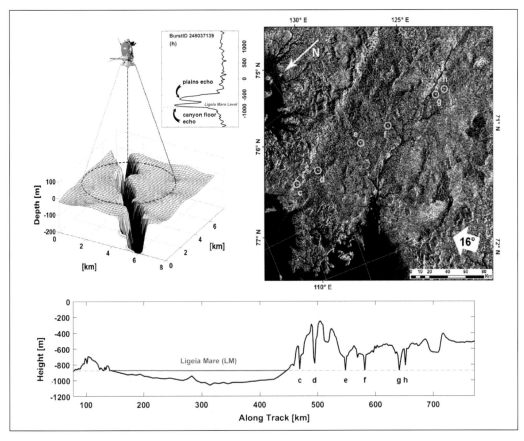

LEFT The river Vid Flumina is seen in the radar image at the upper-left as the dark branching valley network that drains into Ligeia Mare. When the radar was used in altimetry mode (upper-left schematic, red track on the map) echoes from the valley floor (lower pane) were seen at roughly the same level as the sea, suggesting deep canyons.
(Valerio Poggiali)

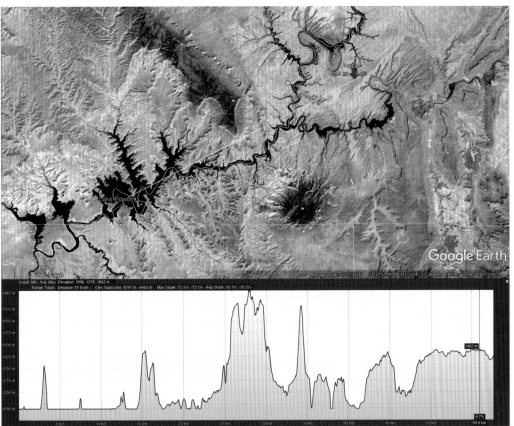

LEFT Flooded canyons of Lake Powell in the southwestern USA have a topographic profile very similar to that measured at Vid Flumina, although the horizontal length scale is a factor of 10 shorter.
(Google Earth)

ROCKS, ICE AND THE STRENGTH OF GEOLOGICAL MATERIALS

The tremendous diversity of planetary bodies challenges traditional (i.e. terrestrial) geological terminology. We think of the solid Earth as being made of rock, with iron in its core and some deposits of ice in its polar regions. 'Rock' generally means an assemblage of minerals such as silicates, carbonates, sulphates and oxides, and 'ice' tends to mean frozen water exclusively.

In the outer solar system, compounds that are gases on Earth can be solids, and are generally referred to as 'ices', but 'water ice' is referred to as such in order to be specific, as there can be CO_2 ice, methane ice, even (on cold Pluto and Triton) nitrogen ice. There are even mixed ices. For example, a frozen aqueous solution of ammonia forms an 'ammonia-water ice' which can possess a range of compositions and properties. Of particular interest is ammonia dihydrate. This has the lowest melting temperature, at only 176K (−97°C), and hence might be the most likely 'cryolava' in volcanic structures on Titan. Another dimension arises from the ability of water to form a 'cage' structure around other molecules; for example, methane. This 'methane clathrate' ice can be found at the high pressures on the terrestrial seabed: mechanically it isn't too different from water ice, but because the methane can be released at sea-level pressure one can set fire to this material. It is likely that Titan's abundant methane was retained during the process of planetary formation by being locked up in clathrate ice. Interestingly, ethane that is percolating in 'groundwater' on Titan might be able to displace methane

BELOW The mechanical stiffness of materials under Titan conditions. Polycyclic aromatic hydrocarbons (e.g. napthalene and phenanthrene), tholins, and water ice are softer, even at low temperatures, than 'soft' rocks on Earth such as gypsum. However, because the environmental collision processes are less energetic on Titan by a comparable factor, the net geomorphological results are similar. *(Xinting Yu)*

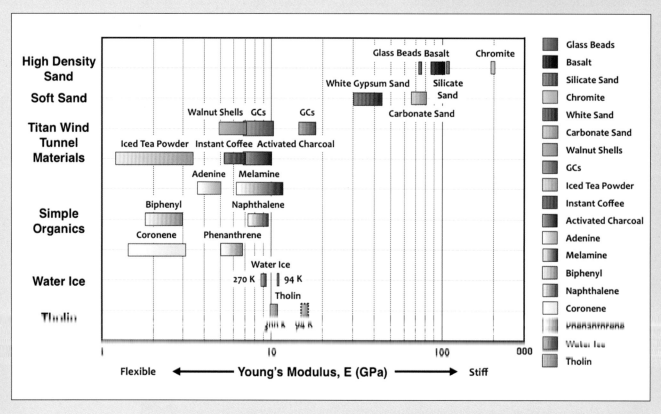

and help replenish methane against destruction in the atmosphere.

Much of Titan's solid material is likely to be organic (meaning carbon-bearing). Some of these organic compounds are gases at terrestrial temperatures (acetylene, butane), some are liquids (benzene, acetonitrile), and some heavier compounds are solids (anthracene).

In the conditions prevailing on Titan all these materials are solid (only methane and ethane are liquid, with propane a borderline case) but can be soft, like candle wax is on Earth. The heavier the molecule, the harder it tends to be. Nitrogen-containing compounds tend to be harder than purely hydrocarbon compounds of the same molecular weight. At 90K acetonitrile is almost as hard as water ice.

Mechanical strength can mean different things – Young's modulus defines how compliant a material is when it behaves elastically; yield strength indicates the stress when a material will permanently deform in a 'plastic' manner; and fracture toughness indicates the conditions in which a material may suddenly crack in a brittle manner.

Although the absolute values of these quantities will be generally lower for materials on Titan, we must remember that the gravity is lower, so rivers flow slowly, and winds are similarly an order of magnitude less than on Earth. The kinetic energy of colliding particles of a given size on Titan is a hundred times smaller, but for Titan's weak materials the net abrasive effect of a single collision may end up being about the same as for the more violent movement of Earth's harder materials.

BELOW The brittle behaviour of typical Titan materials at 90K was assessed by a simple hammer apparatus, inspired by a penetrometer instrument on the Huygens probe. Ice is hard, but fractures. Frozen ammonia dehydrate is less hard and is heavily fractured. Benzene is soft, like wax, and forms a circular dent. Decane is similar. Acetonitrile is the hardest of the simple organics that were tested, with a behaviour intermediate between the dydrocarbons and the water-ice examples. *(Author)*

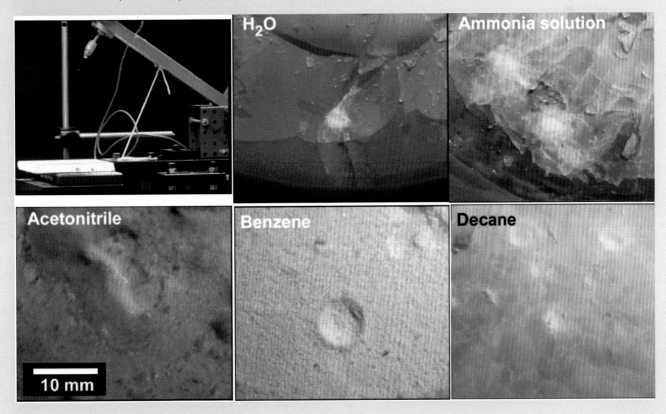

RIGHT A 100x300km section of the T41 radar swath, south of Xanadu, shows complex, dissected terrain, with what appears to be a meandering river channel. The fact the floor is radar-bright suggests coarse gravels and thus an energetic flow. *(Author)*

appear to show erosion down to a lower base level, creating terrains that range from incised plateaus to isolated buttes. To what extent the erosion was mechanical versus by solution is not clear, nor is the reason for the sharp edges of some of these areas.

The most spectacular evidence of river action found by Cassini is associated with the channel Vid Flumina, which connects to Ligeia Mare in the north. In radar images this valley system is seen to be several hundred kilometres long, with several tributaries. But it was also observed by the radar in its altimetry mode, which showed specular reflections suggesting the channel was liquid-filled. Measurements showed that these glints were from a surface that was some 300m lower than the surrounding terrain, implying that the liquid was at the bottom of steep canyons. It should be noted, however, that the glints do not mean liquid was flowing; it was likely back-filled from Ligeia – in much the same way that canyons adjoining Lake Powell in Utah were flooded by the construction of the Glen Canyon dam.

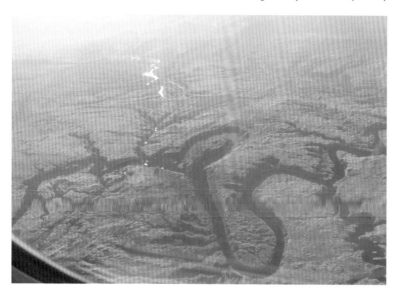

BELOW This view of the Colorado River near Lake Mead shows 'gooseneck' meanders that have incised deeply into rock because a once-sedate shallow slope has steepened by tectonic uplift. There is a hint of a 'bathtub ring' on the steep canyon walls. There is a bright specular reflection off the smooth river surface to the upper-left. *(Airplane window view by author)*

RIGHT This 250x150km scene near the south pole from the T39 swath shows dissected labyrinth terrain as well as dark-floored (thus presumably fine-grained) riverbeds. *(Author)*

Dunes

Even the earliest HST maps of Titan showed dark regions on the trailing hemisphere and it was not unreasonably speculated (like the interpretation of our Moon's dark 'maria' as seas) that these might be lakes or seas of liquid hydrocarbons. There had been radar reflections indicating large millpond-flat areas, but the lack of an optical specular glint (the mirror-like reflection of the Sun on a smooth surface) in both Cassini and telescope data argued against there being open bodies of liquid.

Despite theoretical predictions that dunes would *not* be found on Titan (it was expected to be a rather damp world), the T8 radar observation in the autumn of 2005 showed that the dark area Belet was covered with long, narrow mounds of dark material. They were sand dunes – giant ones, a kilometre or so wide and a couple of kilometres apart, but tens of kilometres long. They were reminiscent of a Zen rock garden, with grooves raked around rocks that stood like islands in a sea of gravel. This 'streamline' pattern implied that the dunes were 'longitudinal', a type of dune that is common in the Namib, Arabian, and Australian deserts on Earth.

Terrestrial aeolian geomorphologists ('dune whisperers') had recognised that this type of dune arises when the wind tends to occur in two different (converging) directions. As sand is pushed one way and then the other, it tends to accumulate in a long, narrow and somewhat symmetrical mound that lies roughly along the mean wind direction. This is distinct from what is seen when the wind direction is uniform and there is abundant sand. That produces an asymmetric set of ridges at right angles to the wind that possess the steep avalanching slip faces in the downwind direction. When the sand supply is limited, these transverse dunes break up into curved ridges, and ultimately into isolated crescent-shaped 'barchans'. On the other hand if the wind blows in multiple directions, the sand can be swept up into isolated star-shaped dunes.

The term 'sand' in geology refers to the size and nature of the particles, and does not signify a similar composition to the sands of Earth or Mars. Titan's sand was optically dark, suggesting that it might be organic materials rather than ice. It also had microwave

GEOLOGICAL PARTICLE SIZES

To geologists the term 'sand' just means particles with diameters in the range 0.05mm to 2mm, without considering their composition. Most terrestrial sand is made from the mineral quartz, but there are volcanic black sands of basalt, white sands of gypsum, and even purple sands of garnet. Titan's sand appears to be dark. It is probably carbon-rich material, perhaps a mix of polycyclic aromatic hydrocarbons (PAH).

The grain diameters for different geological classes of material are as follows:

Name	Grain diameter
Boulder	>256mm
Cobble	64–256mm
Gravel	2–64mm
Granule	2–4mm
Sand	1/16–2mm
Silt	1/256–1/16mm
Clay	<1/256mm

Note that various grades exist for sand and gravel: fine, medium, coarse, etc.

The rounded 'rocks' at the Huygens landing site were interpreted as being water ice, but the possibility that they were of some organic composition cannot be ruled out.

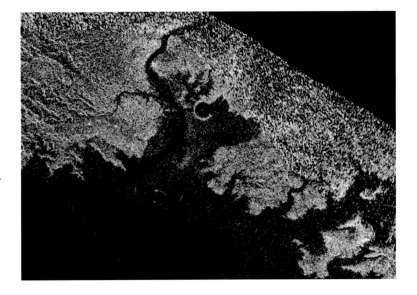

BELOW A radar image of a river (Xanthus Flumen) that seems to drain into Puget Sinus at the northern margin of Ligeia Mare. The stippled pattern suggests that silt deposition has made the sea shallow in this inlet, permitting some brightness from a bottom reflection. Nevertheless, the dark arcing path of the deeper river channel can be traced. A small, nearly perfectly circular arc could be the vestige of an impact crater. The image is 200km across. *(Author)*

RIGHT The October 2005 radar image (top) revealed the true nature of Titan's equatorial dark areas as sand seas. The long parallel ridges apparently flow like streamlines around bright mountain 'islands', reminiscent of a Zen rock garden. *(Author)*

LEFT An image by an astronaut on the STS-107 mission looking obliquely across the Namib Sand Sea toward the Atlantic Ocean. These parallel ridges are among Earth's largest sand dunes, and have a bright/dark appearance due to the relatively low Sun angle forming shadows toward the lower left. It was this image that helped us to recognise the nature of the dunes in the Cassini radar images because they have the same shape and size. *(NASA)*

properties (sensed by the radar) consistent with organics but not with rock, which was not in any case expected to be exposed at the surface of Titan. Some additional hints on its composition came from Cassini's VIMS, which showed that the dark dune-covered areas had a small absorption band near 5μm that was characteristic of 'aromatic' organic chemicals (those with one or more rings of carbon atoms).

One of the most striking fields of these longitudinal dunes on Earth is found near the coast of Namibia. They can be over 100m high, and extend in long lines running roughly north–south, sculpted by winds that change direction with the seasons.

The discovery of such dunes on Titan stimulated extensive work on how dune shape relates to environmental parameters. Radar and infrared measurements showed that the

LEFT A view from a few hundred metres' altitude of the north end of the Namib Sand Sea looking south, showing a perspective like that from the Dragonfly vehicle planned for the mid-2030s. While the dune crests are steep, wide interdune plains offer abundant safe landing sites. *(Author)*

RIGHT A Space Shuttle radar image of the eastern edge of the Namib, with a resolution ten times better than that of Titan from Cassini. It shows the contrast between bright mountains and intermediate gravels of the Tsondab river, and the dark sand. The glints highlight that the dunes have a complex structure. The landscape is a continual 'tug of war' between sand being blown in growing dunes and swept away by occasional rivers. *(NASA/JPL/DLR/ASI)*

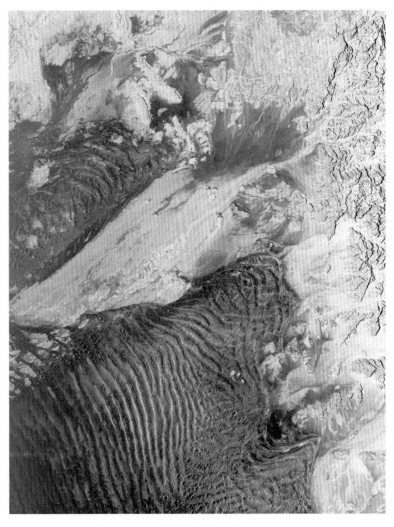

dunes on Titan reached heights of 100–150m in places, so their shape and size was essentially the same as those in the Namib or Arabian deserts. This similarity was a puzzle, given that the sand was made of different stuff, the gravity is a factor of 6 different, the air 4 times denser, and so on.

Now, Namibia, desert streaming into ocean, waves of bright sand diving into dunes of dark water
– visible rhythms of blue and brown, sea and sand dance upon my strings.

Astronaut Story Musgrave, 9 June 1999

LEFT A handheld camera image (500mm focal length) from the International Space Station of the Tsondab river valley and the nearby dunes (see the accompanying radar image). Notice the complex dune shapes. Such details are too small to be visible in a Cassini image. *(NASA JSC/EOL)*

ABOVE A Cassini view of dunes on Titan diverting around a rather elephantine radar-bright topographic obstacle. *(Author)*

ABOVE Sand dunes near Alice Springs, Australia, being diverted by, and occasionally penetrating through a set of low hills. *(Airplane window view by author)*

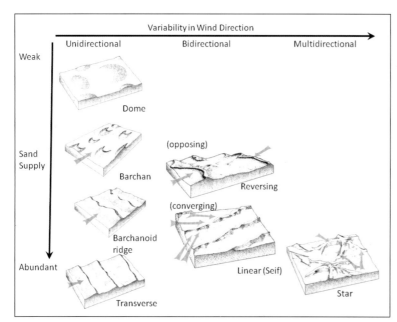

LEFT The shape of a dune correlates with the diversity of the wind. Titan's dunes are almost exclusively of the linear type, indicating a wind that predominantly alternates between two directions. *(USGS)*

Titan's dunes proved to be important 'wind vanes' that indicate the long-term behaviour of the climate. Their arrangement is particularly important because Titan has so few clouds that can be tracked.

The first thing to note is that the very existence of the dunes indicates that sand-sized organic material is available, is sometimes dry, and the wind is at least occasionally strong enough to blow it about.

Secondly, the dunes are in a single globe-circling belt that extends up to about 25° either side of the equator. This is in contrast to Earth, where the deserts form two belts, about 20°S and 20°N of the equator. The controlling factor here is Titan's slow rotation. It has the effect of drying out the low latitudes. On the fast-spinning Earth, atmospheric circulation

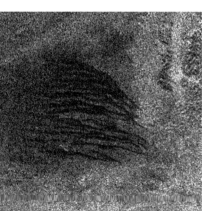
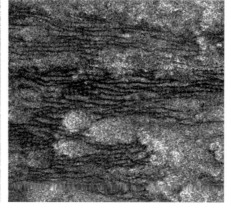

LEFT These dunes west of Menrva have a tapering shape, suggesting a limited sand supply. The forked arrangements are indicative of a recent change of wind regime. *(Author)*

ABOVE One puzzling feature of Titan's dunefields is how abrupt their edges can be. It may be that ephemeral rivers, too small to clearly observe in the Cassini data, sweep the sand away faster than the dunes bring it. This happens at the Kuiseb river, at the northern margin of the Namib Sand Sea, seen in an image from the Space Station. *(NASA)*

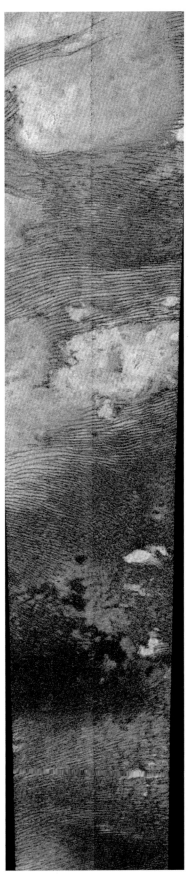

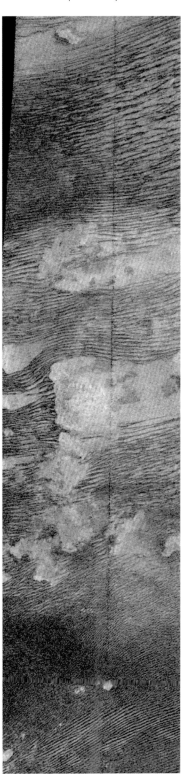

LEFT The T92 radar swath showing lots of dunes. *(NASA/JPL)*

BELOW The T55 radar swath showing lots of dunes. *(NASA/JPL)*

produces a downwelling of dry air in the two mid-latitude belts.

The shape of the dunes on Titan showed that there must be two predominant wind directions. Given that Saturn takes almost 30 years to travel around the Sun, the very long seasons create an equatorial near-surface flow on Titan that is predominantly northward in that hemisphere's summer, and southward in northern winter.

The exclusively eastward orientation (with a few regional variations) of the dunes was initially difficult to understand. On a rotating planet, the near-surface flow at the equator ought to be to the west, just like our trade winds that let sailors reach the Americas from Europe or the ocean currents that drift from the mid-Atlantic toward the Gulf of Mexico. The vital point, however, which took some time to unravel, is that dunes do not record the average wind direction, only the integrated effect of the winds that have sufficient strength to move sand.

Although the near-surface equatorial winds on Titan, flicking north and south with the seasons, have a westward bias, the winds

RIGHT A map of dune orientation circa 2008, indicating the predominantly eastward orientation (confirmed by more extensive subsequent mapping). Notice that the ISS basemap just shows Kraken at the upper-left, but has not covered the north polar regions because those were still in darkness. *(Author)*

further aloft move in the prograde sense of planetary rotation, namely west to east. Just as occurs on Earth, it may require rather blustery conditions to move sand, and on Titan these conditions apparently occur near the equinox, when the Sun is nearly overhead. Methane rainstorms (or perhaps just convection more generally) diminish some of that eastward momentum, such that the wind blows to the north-northeast or south-southeast more often than the westward counterpart directions. As a result, the dunes are assembled in a longitudinal manner with an eastward growth direction.

The spacing of the dunes is another diagnostic of the atmosphere. Dunes with avalanching slip faces (as opposed to ripples, which form a slightly different way) can begin to form whenever a sand patch is larger than a certain size (on Earth this is a few metres, but on Titan it is several tens of centimetres). As sand accumulates, the dune will grow in width and height until it gets so tall that the airflow is squeezed at the top, where the accelerated airflow removes sand from the crest as fast as it is delivered there. The squeeze occurs when the dune approaches a height of about 10% of the thickness of the convective boundary layer, the part of the atmosphere that is warmed directly by the ground (often this layer is evident on Earth from a height of several hundred metres up to several kilometres as a layer polluted by smog or other haze). As a result, depending on the supply of sand and the age of the dune system, the height of the dunes will range from a few tens of centimetres to about 100–200m, and because the aspect ratio of height to spacing is about 1:20, regular dune systems can have a spacing up to several kilometres. The boundary layer is generally thinner near the sea, where the thermal inertia of the ocean acts to prevent the air warming up as effectively, so the dunes at the edge of the Namib Sand Sea are smaller than those in the middle. On Titan, where the seas are not present near the equatorial dunefields (being instead in the polar regions), the boundary layer is everywhere ~3km thick, and the dune spacing is remarkably uniform planetwide.

Finally, the very regularity of the dune pattern on Titan tells us something. If the wind regime changes substantially, the dune pattern must re-align too, although doing so will take a while. For large dunes 100m high, this reorientation timescale (approximately the same as it takes to build a dune from scratch) is tens of thousands of years for both Earth or Titan. And while the pattern is adjusting, we see a mix of the old and the new.

There are a handful of places on Titan where

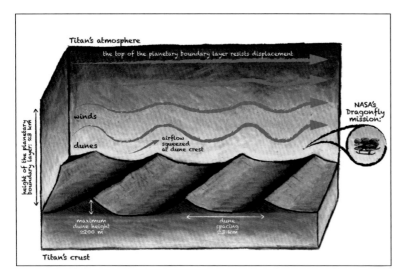

BELOW The top of the planetary boundary layer (PBL) resists upward movement, and thus compresses the airflow at the dune crest. As a result, the dunes only grow to a height of about 1/12th of the PBL height. Because dunes tend to have a width ten times their height, the observed spacing for mature (fully developed) dunes reflects the PBL height. On Titan, this appears to be a rather uniform 3km or so. *(James Tuttle Keane)*

the uniform longitudinal mould is broken, and a superposed pattern of apparently transverse dunes is seen. But this only happens adjacent to a bright region, where a topographic obstacle interrupts the global circulation and forces a local straightening of the flow.

In some places on Earth we can trace the source of sand – a good example being White Sands National Monument in New Mexico, where brilliant white gypsum sand is blowing off a dry lakebed (Lake Lucero) and making a field of dunes that extends for several tens of kilometres downwind. In contrast, in the Sahara (and similar locations) the sand has been swept around and around for an extended period.

On Titan, we have not yet been able to identify the sand source. The dunefields form a nearly unbroken belt around the equator. It is not known whether the winds divert sand around the notable obstruction of Xanadu or whether rivers play a more significant role. Perhaps the sand forms by haze particles falling into lakes and seas, which subsequently become desiccated. The haze particles, now washed clean of more-soluble organics, will clump together to create sand grains which are then blown off the seabed. Of course, they will have to migrate thousands of kilometres equatorward (which is uphill). Perhaps the sand forms by the processing of surface organics by the small fraction of cosmic rays that reach the surface. Only when we are able to examine the detailed composition of the material itself will we gain further clues.

The sand has to be hard enough to survive transport over thousands of kilometres. However, that may not be so difficult, given that the energy involved is rather low. That is, small grains moving at ~1m/s do not have much kinetic energy. As it happens, the breakdown of organic granulated materials by transport in an airstream is an important design consideration in the food and pharmaceutical industries, where pneumatic conveyance in pipes is an economical technique.

But seas of sand, beautiful though they may be, are not as evocative as the seas of actual liquid that scientists were hoping to find on Titan. It was only when the north polar region hove into view halfway through Cassini's initial four-year tour of the Saturnian system that these showed up.

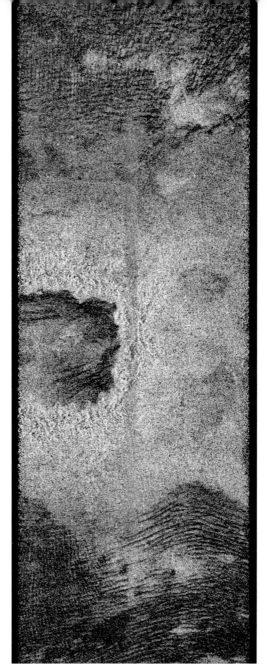

LEFT The T95 radar image of the Selk impact crater, ~70km in diameter. Dragonfly may land in the arching field of dunes to the southeast. Note some dunes on the crater floor and the presence of north–south-aligned dunes that are superposed on the generally east–west pattern. These may resemble the 'raked' dunes seen in the Namib. *(Author)*

BELOW A view over the Namib from a height of a few hundred metres (probably typical of a Dragonfly flight), showing the near-orthogonal 'raked' pattern seen occasionally on Titan. *(Author)*

Chapter Six

Lakes and seas

Other than Earth, Titan is the only world in the solar system with open bodies of liquid on its surface. These lakes and seas, filled with liquid methane and other organics, are both familiar and exotic.

OPPOSITE A radar mosaic of Ligeia Mare, Titan's second-largest sea, showing many features common to lakes and seas on Earth. The image is about 400km across. The calm sea surface appears pitch-black, except near some edges where some bottom reflection may be detected. The shoreline betrays a complex geologic history, with drowned valleys indicating a recent rise in sea level. The rectilinear arrangement of channels at the bottom-left suggests some tectonic control on hydrology. *(Author/Cassini Radar Team)*

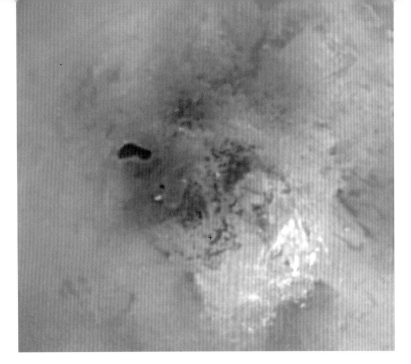

ABOVE An early Cassini ISS (near-infrared) mosaic, including the kidney-shaped feature that would later be named Ontario Lacus. At the time, however, the myriad of unfamiliar bright and dark shapes on Titan made it difficult to determine what we were seeing – and speculations included a sinkhole filled with dark solids. Note the faint, bright-edge 'bathtub ring'. The south pole is marked by a red cross below-centre. The brightest features seen here are methane clouds. *(NASA/JPL/SSI/U.Arizona)*

BELOW In 2006 the first radar swath (T16) of high latitudes on Titan showed a plethora of small lakes. Many had bright rims, the origin of which is still not known. The large lake at lower left was later named Bolsema Lacus: others here include Koitore and Mackay Laci. *(Author/Cassini Radar team)*

Methane and ethane are both liquids at the temperatures on Titan, therefore it was natural to speculate that there might be lakes or seas. However, for the first couple of years of Cassini's exploration open bodies of liquid were elusive. There were optically dark lake-shaped patches at the south pole and optically and radar-dark regions girdling the equator, lapping against the edges of bright highlands, but the latter turned out to be seas of sand. The damp ground at the Huygens landing site attested to moisture, and there had been rivers at least in the past, there and elsewhere, but for aesthetic reasons as much as any, scientists wondered if there might be open seas.

It turned out that the polar regions favoured the accumulation of liquids; not only were they a little cooler, they were also lower in elevation. The first undisputed indication of lakes was on flyby T16 in the summer of 2006, the first high-latitude radar swath – although the area under investigation was in winter darkness. This showed dozens of radar-dark lakes, together with lake-shaped pits that seemed more abundant toward lower latitudes. Some of the lakes were connected by fine dark channels. The radar reflectivity of the dark ('filled') lakes implied that they were either liquid hydrocarbons (i.e. methane and/or ethane) or possibly fluffy organic dust which had somehow filled pits and channels that had been previously carved by liquid. Despite the failure of Occam's razor in the past on Titan, the straightforward explanation of present-day liquids was borne out.

In the following year, larger expanses of the north polar region were imaged, revealing more small lakes, typically 10–20km wide, and several much larger bodies of liquid were mapped. Cassini's optical instruments also got to study closely the only large lake in Titan's southern hemisphere – the 250×70km lake that came to be called Ontario Lacus. (It was decided that lakes on Titan would be named after terrestrial lakes – which so far has not led to too much confusion – and this one, being the size and shape of Lake Ontario, got that Latinised name.) Although it is challenging to extract spectroscopic information about the surface of Titan, in particular methane and ethane, while peering through a deep

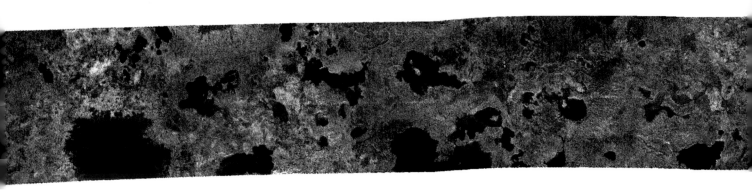

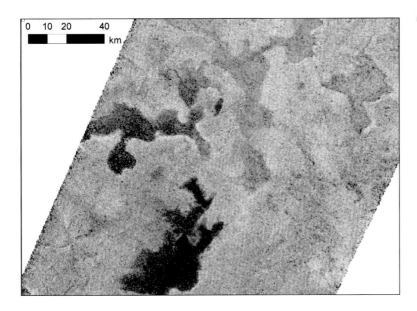

ABOVE A close-up of some of Titan's lakes, and some nearby lake-shaped depressions that appear not to be liquid-filled at present. *(Author/Cassini Radar team)*

BELOW Although radar data show that Titan's lakes sit in depressions, the rims of many of the small lakes are raised, like ramparts. Perhaps basins are able to erupt material that settles on the rim. *(NASA/JPL artist's impression)*

There are three seas on Titan, each the size of Earthling Lake Michigan. The waters of all three are fresh and emerald clear. … There is a cluster of ninety-three ponds and lakes, incipiently a fourth lake. … Connecting the [seas and pools] are three great rivers. These rivers, with their tributaries, are moody – variously roaring, listless, and torn. Their moods are determined by the wildly fluctuating tugs of eight fellow moons, and by the prodigious influence of Saturn.

Kurt Vonnegut, *The Sirens of Titan*, 1959

BELOW The author on a TV shoot on Lake Powell, straddling the Utah–Arizona border in the southwestern USA. The 'bathtub ring' of light-coloured evaporites is glaringly apparent on the steep walls of the canyon. *(Author)*

ABOVE A colour overlay of VIMS data on a radar image, showing the distinct red (5μm bright) deposit at the margin of Ontario Lacus, probably a 'bathtub ring'. *(Shannon MacKenzie)*

atmosphere that is itself laden with methane, the Cassini VIMS instrument was able to verify the presence of ethane in that lake. There was also a hint that it was muddy. Also, its margins were bright, suggestive of a 'bathtub ring' deposit of formerly dissolved materials (analogous to salts in terrestrial lakes).

The shape of Ontario resembled not only its namesake, but also a closer geomorphological analogue: Racetrack Playa in Death Valley (some scientists also pointed to Etosha in Africa). Racetrack is bone-dry for 99% of the time, and is incredibly flat, such that only a couple of inches of rain is enough to flood it completely. Cassini's radar, operating in altimetry mode, found that the margins of the lake had slopes of only about 1:10,000, and the lake itself was mirror-smooth at scales of several hundred metres. In fact, it was such a perfect mirror that the dazzling reflection saturated the radar receiver. It was determined that if there were any waves or ripples on its surface, they had to be less than a couple of millimetres in amplitude.

Radar imaging showed one margin of Ontario Lacus to be rather straight, possibly wave-cut, and the eastern side had a 'rabbit-ear' delta at the mouth of a narrow, winding river channel, showing the river had deposited sediment in the lake. The radar also showed the margins of Ontario as not being pitch-black like most of the northern lakes, but what might be a bottom reflection through a shallow layer of partly radar-transparent liquid. If the slopes within the lake were anything like those at the margins, then even tens of kilometres from the shoreline there would only be a few metres of liquid, a situation akin to the turquoise margins of desert islands, so a contribution to the radar brightness by a bottom reflection wasn't unreasonable.

Although there are always challenges in comparing data collected by different instruments, there were some indications that the outline of Ontario in the 2009 radar image was different from that in the 2004 near-infrared one – a uniform 'smudging' of the latter outline might be expected because of scattering in the atmosphere, but there were parts of the radar edge that were substantially inward from the rest. If the slope of the lakebed were only

BELOW An aerial view of Racetrack Playa in Death Valley, famous for its 'sliding rocks' (which this author has seen in motion). Although 50 times smaller than Ontario Lacus, there are morphological similarities, particularly alluvial deposits to the upper-right and a straighter (possibly wave-cut) margin on the right. Coincidentally, mountains abut both lakes at the south (bottom-right). *(Author)*

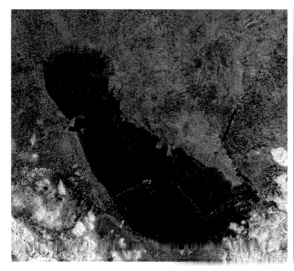

1:10,000, a real change in lake outline might only require about 1m/year of evaporation, which was not very different from the prediction.

Such a change (if it is real) would likely be a seasonal effect. The paradigm of a drying-out of Ontario was implied by its 'bathtub ring', its shallowness, and its shape. This was consistent with a broader mystery that had emerged, namely the gross difference between the north and south hemispheres. Whereas the south had only Ontario and a handful of other small and/or shallow features, the north was replete with hundreds of small lakes and three

LEFT A Space Shuttle image of the 25x56km Salton Sea in southern California, a lake created by accident(!) in 1905 that is progressively drying out and has obvious morphological similarities with Ontario Lacus. Note the deltaic promontories at the north and south, and the dry riverbeds and alluvial fans that drain into the lake at the right and left. *(NASA)*

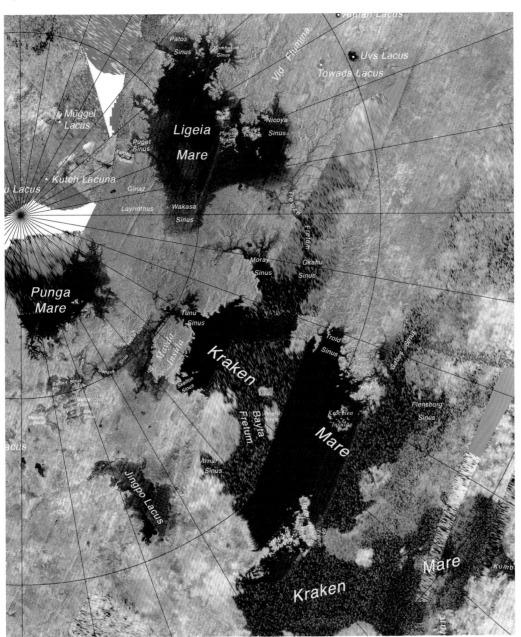

LEFT Titan's three seas and many lakes, identifying the principal named features. This is a radar mosaic with an artificial black-blue-brown colour scale to guide the eye. Blue areas are likely shallow, with a bottom reflection contributing to the brightness of the image, although in some cases the speckled blue appearance is due to poor signal. *(USGS)*

that were large enough to merit designation as 'seas'. Spanning 1,000, 400 and 250km, these were named after the sea monsters Kraken, Ligeia and Punga, respectively.

Altogether, the seas covered about 1% of Titan's surface globally, or about 10% of the surface above latitude 55°N. But there was nothing about the terrain to suggest why the north should be so endowed with hydrological abundance while the south was so deprived. The two polar regions were at an elevation ~0.5km lower than the equator, so the difference had to be in the atmosphere, the key being the configuration of the seasons.

Saturn's orbit around the Sun is eccentric, so one pair of seasons is longer than the other. As it happens, at present, southern summer is shorter but more intense than the north. As a result, the peak temperatures on Titan in the south are higher and 'cook off' volatile materials such as methane and ethane, distilling them over to the cooler north. It is the polar summer when most rainfall occurs, and the rainy season is longer in the north. The result is that the north has been accumulating liquid as the south dries out. This has made Ontario Lacus ethane-rich and given it a margin of evaporites.

The interesting corollary of this explanation – which can be reproduced in models – is that this astronomical configuration of the seasons, just like that on Mars and Earth, isn't constant with time; it changes predictably over a period of tens of thousands to millions of years. Specifically, the situation was reversed ~30,000 years ago, with the northern hemisphere having the shorter, hotter summer. And presumably then the southern seas were filled, and the north was drying out. The corresponding evolution of seasons on Mars leads to the accumulation of alternately dust-rich and ice-rich material in its polar caps, giving rise to spectacular layered deposits (in which there is a fascinating record of Martian climate change). On Earth, these changes drive our Ice Ages. So we see that the same physical process drives profound climate changes on all three worlds, albeit with different manifestations.

The dynamics of the redistribution of volatiles on Titan leads to some other things that were detected by Cassini. Close inspection of the area around Ontario Lacus revealed vestiges of a shoreline of a basin much larger than the lake itself and its bright evaporite rim. Presumably this basin was substantially filled ~40,000 years ago, a time when woolly mammoths roamed northern Europe and the last Neanderthals were giving way to modern humans.

The margins of the seas in the northern hemisphere of Titan, notably the southern part of Ligeia Mare, has an intricate 'fingered' coastline, with branching island chains and narrow inlets. This arrangement, called a Ria coastline, is characteristic of rising sea level (or land sinking relative to sea level). The island

RIGHT Brilliant sunglint (the yellow spot at about 11 o'clock) is seen in this VIMS composite near Seldon Fretum of Kraken Mare, with the Sun about 40° above the horizon. Punga Mare is visible just above the bright arrow-shaped set of clouds over Ligeia. Paler orange evaporite deposits are seen especially around the margins of Kraken. (NASA/JPL-Caltech/U. Arizona/U. Idaho)

RIGHT Ontario Lacus in a wider basin, with an interpreted paleoshoreline indicated in red. This former shoreline is about 200m higher in elevation than the lake's present boundary, which is at an elevation of −826m. This basin, along with three others in the south, could easily accommodate the amount of liquid presently seen in the north.
(Cassini Radar Team)

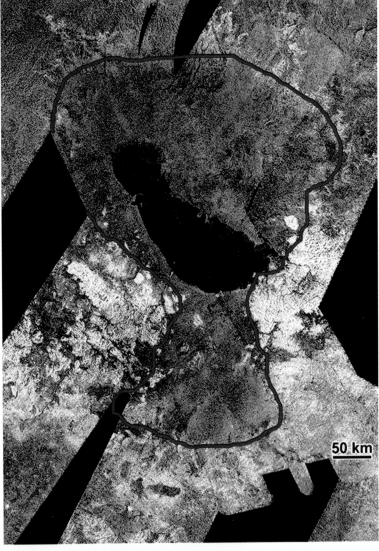

chains are ridges, and the inlets former river valleys. Obviously a river valley can only be carved efficiently when it is exposed, and so this morphology is evidence of these valleys being flooded by the sea. Some good terrestrial examples are Lake Mead and Lake Powell in the southwestern USA (the recent flooding being the result of damming the Colorado river). Another good example is found in northern Oman, where the Musandam mountains are sinking into the Arabian Gulf as the Arabian tectonic plate is subducted beneath that of Eurasia. Identical morphologies at Ligeia are consistent with it filling up at present.

One of the most remarkable observations by Cassini was in 2013, when the radar was used in altimetry mode, looking straight down into a sea. Not only did this observation confirm that Ligeia at that time was flat as a millpond, as Ontario had been, but about 1 microsecond after the dazzling echo resulting from the mirror-flat sea surface, a second echo could be discerned, about 1,000

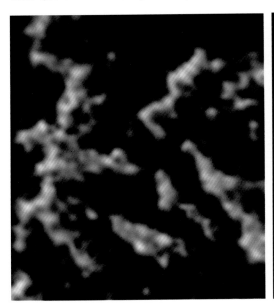

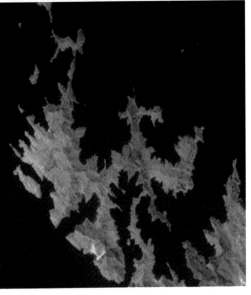

LEFT At the left, the narrow branching islands of Planctae Insulae in Ligeia Mare are indicative of ridges carved by erosion, where the sea level has subsequently risen and flooded the intervening valleys. At the right is an optical satellite image of the Musandam Peninsula in Oman at the same scale, showing identical morphology owing to the sinking of the mountains into the Arabian Gulf.
(Author, with NASA image data)

PHYSICAL PROPERTIES OF LIQUID METHANE AND ETHANE

There are in fact two compounds that are liquid at Titan temperatures – methane and ethane. Technically propane may be liquid too, but is not very abundant. Other compounds, notably nitrogen, are not liquids by themselves at 94K, but can dissolve in liquid methane or ethane. Although both are components of 'natural gas' on Earth, they have quite different properties. Liquid methane is less than half as dense as water, ethane is about two-thirds as dense.

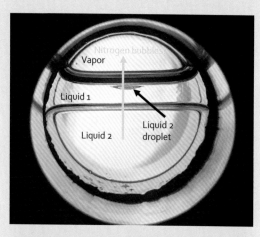

RIGHT Laboratory experiments on liquid ethane/methane/nitrogen mixtures indicate that at certain pressures and temperatures, two different liquid phases can co-exist. *(Jennifer Hanley/Lowell Observatory)*

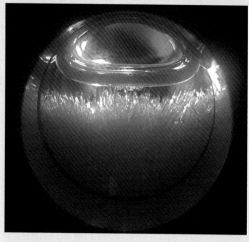

RIGHT A 60% ethane and 40% methane mixture being frozen. Ice crystals form on the bottom of the chamber. *(Jennifer Hanley/Lowell Observatory)*

Other physical properties (including some that the Huygens probe was intended to measure in the event of a splashdown) such as dielectric constant, refractive index, or speed of sound, have similar differences, and measurements could serve as a simple way of estimating the ratio of methane to ethane in the sea.

A bigger difference is viscosity. While methane is about 5 times less viscous than water, ethane is about double (and if propane and heavier compounds were dissolved in ethane, the solution would be even more gooey). However, the most profound difference is their volatility. At 94K the saturation vapour pressure of methane is about 0.17 bar, so 100%-saturated air on Titan is about 11% methane. In contrast, ethane has a pressure of only a few microbars. In comparison, water at 0–20°C on Earth has a vapour pressure of 0.006–0.02 bar, so 100%-saturated air is 0.6–2% water. On Earth, ethane is involatile, much like engine oil – it may evaporate over long (geological) timescales but as far as human perception and meteorology are concerned, ethane liquid just sits there.

The density of ethane is ~⅔ that of water and the viscosity at 94K is rather similar, depending on temperature. Methane is a little less dense (about half that of water) and rather less viscous. If dissolved constituents such as higher hydrocarbons and nitriles are also present, they would increase the density, viscosity, and dielectric constant. Surface conditions on Titan likely don't permit methane and ethane to freeze, due to the antifreeze effect of dissolved nitrogen. Unlike water, solid hydrocarbons are generally denser than their liquids, so we should not expect any 'icebergs'.

TABLE: PROPERTIES OF LIQUIDS AT 293K (EARTH) AND 94K (TITAN)

Fluid	Water	Methane	Ethane
Density (kg/m^3)	1000	450	650
Viscosity (μPa·s)	1000	200	2000
Surface Tension (N/m)	0.073	0.018	0.018
Sound Speed (m/s)	1482	1574	2000
Refractive Index	1.33	1.287	1.38
Dielectric Constant	~50	1.65	1.9
Thermal Conductivity (W/m/K)	0.6	0.2	0.25
Specific Heat (J/K·kg)	3.3	2.2	4.2

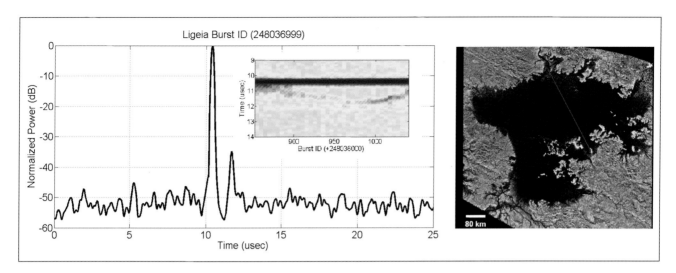

ABOVE In 2013, Cassini's radar in altimetry mode stared into the abyss of Ligeia Mare. The strong surface echo showed that waves were not present, which also meant there was no clutter to obscure a second echo, from the seabed, about 2μs later. This second echo tracked deeper toward the centre of Ligeia and shallow at the edges, with a maximum depth of 160m.
(M. Mastroguiseppe/Cassini Radar Team)

times weaker. A distinct reflection from the seabed, this second echo was closer to the surface at the periphery of Ligeia than it was in the middle. At the speed of light in liquid hydrocarbon (about a factor of two slower than in vacuum) a microsecond corresponds to the two-way travel time through ~150m of liquid, making this the first example of extraterrestrial depth-sounding! This detection was possible because Ligeia was not only sufficiently deep but also unexpectedly transparent to microwaves. Ten times shallower and the bottom echo would have merged undistinguishably into the surface echo; ten times deeper (and/or with liquid ten times more absorbing of microwaves) and the bottom echo would have been attenuated away. This observation prompted further attempts at depth sounding in the northern lakes and seas during Cassini's extended mission, as well as a re-examination of Ontario.

New laboratory measurements were made to understand the properties of liquid methane in these extraordinary circumstances. Instrumentation in liquefied natural gas tanks relies upon these properties, but it has never had to confront propagating through hundreds of metres of liquid. Knowing that the signal had to go through ~300m of liquid, the amplitude of the echo required the liquid to be almost free of ethane, propane or any of the other heavier and more absorbing compounds. It is therefore a rather pure methane–nitrogen mix (nitrogen dissolves quite efficiently in methane, resulting in some interesting possibilities). In effect, Ligeia was 'fresh water',

clean methane distilled out of the atmosphere as rain. Later analyses (requiring advanced signal processing to correct for the dazzling) gave some insights about Ontario – it reached a depth of a few tens of metres but was almost half ethane, a rather more absorbing composition.

When the depth-sounding experiment was tried on Titan's largest and most equatorward sea, Kraken Mare, no bottom echo could be detected apart from in a shallow inlet on its northern coast called Moray Sinus (as with lakes, inlets on Titan were named after terrestrial inlets). If Kraken's composition and seabed were similar to Ligeia it should have been possible to see an echo for hundreds of kilometres of the radar track. There are several possible explanations for its absence: the liquid may be more absorbing than Ligeia, the seabed somehow less reflective, or the depth excessive (or some combination thereof). The margins of Kraken in this area are rugged, with terrain rising hundreds of metres above sea level within a few kilometres of the shore, so it would be plausible for the seabed to fall away steeply. There is nothing to suggest that Kraken's seabed should be less reflective, although some graded layers of thin mud

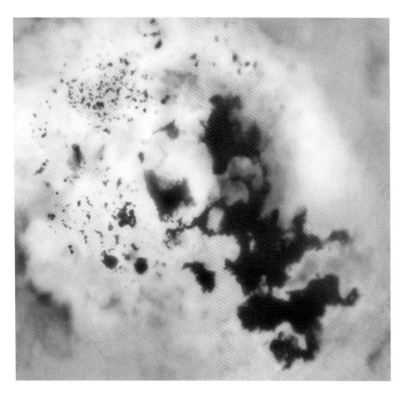

ABOVE An ISS mosaic of the northern polar region, showing the system of major seas at the right, and the plethora of small lakes at the left. A dark shading around Ligeia seen here is not evident in radar data. This mosaic shows well the (tectonic?) straight edges of Punga Mare. *(USGS)*

could serve as an antireflective coating (rather like 'stealth' paint).

To me, the most probable explanation for Kraken's hidden depth is a more absorbing liquid composition, even though Ligeia and Kraken are connected (by the labyrinthine channels of Trevize Fretum). Models suggest rainfall is more extensive at the highest latitudes (a picture borne out by observations of clouds, at least), meaning 'fresh' methane-rich liquid is likely in the seas Punga and Ligeia, nearest the pole and their watersheds. Like the Baltic Sea or Black Sea on Earth, both of which are fed by fresh rivers, the gradient in precipitation could drive a net flux of liquid from Ligeia into Kraken that continuously flushes Ligeia liquid – including whatever ethane and propane is dissolved – into Kraken. The methane is able to evaporate at any point, but it will do so particularly at Kraken's lower latitude, where it is warmer. There, the less volatile solutes like ethane accumulate, rather like salt does in the Mediterranean Sea.

This picture, with its just-like-Earth-only-different cycle, seems to explain the observations, although it can't be proven without measurements of the composition of Ligeia and Kraken. Perhaps it will be possible to sweat some constraints out of Cassini's vast trove of data, but more likely we will have to await a future high-precision orbital survey, or even splashing a pair of probes into the two seas or landing in one sea and then sailing through Trevize to the other.

In principle, if precipitation were the only process forcing the transport of liquid, this would lead to a complete purging of solutes into Kraken, with only fresh rainwater (methane/nitrogen) in Ligeia. But in practice there is some backwash – tides will exchange some salty Kraken liquid with Ligeia on each orbit – models suggest several cubic kilometres of liquid may pass between the two seas on each cycle. The steady-state composition depends upon how vigorously this remixing takes place against the precipitation flux. This could be determined by quantitative composition measurements.

On Earth, some of this remixing occurs continuously. Hence, the surface current through the Straits of Gibraltar is as the precipitation balance would suggest, from the fresher Atlantic into the warm evaporative basin of the Mediterranean. However, 100m below the surface there is a counterflow of dense, salty water. Titan's hydrology could easily have similar complications.

LEFT A schematic of the possible exchanges of liquid between the main sea basins on Titan. The fluxes through the labyrinth (later named Trevize Fretum) and the Throat of Kraken (Seldon Fretum) have net and per the components. P and E refer respectively to the precipitation fluxes into and evaporation out of each basin. *(Author)*

RIGHT Punga and Ligeia Maria may be analogous to the Black Sea, in being kept 'fresh' by rivers bringing methane-nitrogen, and flushing their 'salts' (ethane and other organics) into Kraken Mare (analogous to the Mediterranean), where warmer temperatures cause a net loss of methane by evaporation that concentrates the less volatile solutes. *(NASA)*

BELOW Possible seasonal variations in lake levels on Titan turn out to be rather small. Notice that the precipitation flux into each sea occurs in summer (L_s=90° is northern summer equinox; Ontario in the south gets its rain at L_s=270°). The models assume different ground porosities and 'methane tables'; i.e. how deep you'd need to drill a well on Titan. *(T. Tokano)*

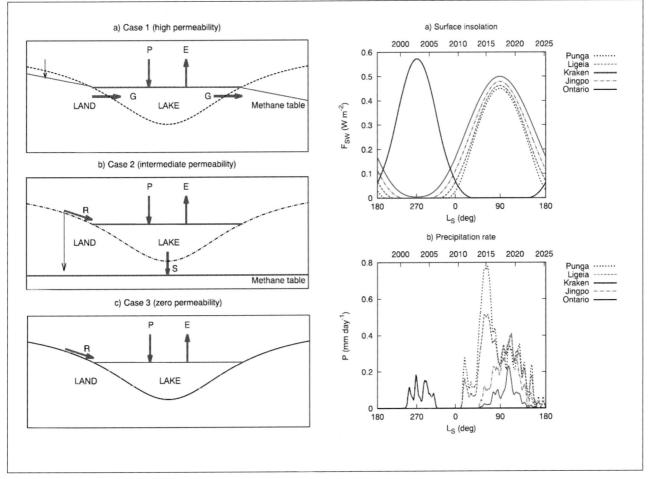

109

LAKES AND SEAS

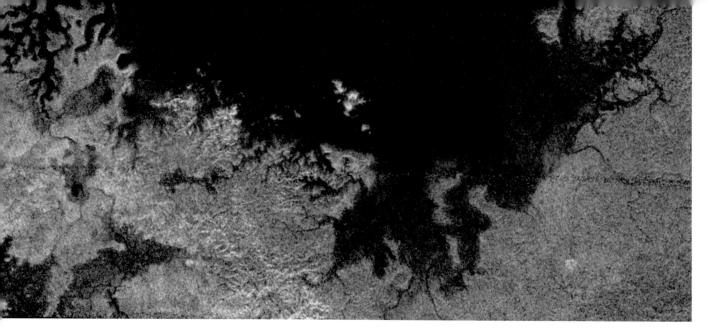

ABOVE A part of the T29 radar swath featuring Punga Mare, the smallest of the three seas. The shoreline morphology is very complex, betraying a range of coastal processes. Close inspection of the full set of radar images suggests Punga connects to Kraken by way of a very narrow channel. *(Author)*

The Kraken–Ligeia connection is probably the most interesting one, but close examination of radar images shows that Punga is also connected to Kraken, and Cassini altimetry shows that the sea level is the same in each, as one would expect if they were hydraulically linked (in fact, they show that the level is very slightly different, but just as much as the tidal potential would predict; see box opposite).

Circulation in Titan's seas might be driven by solar heating, if the seas are transparent enough to let heat be deposited at their base. But this effect may be difficult to isolate from circulation driven by wind stress on the sea surface, which will also vary seasonally. In both cases there will be a strong seasonal variation, whereas tides will show an essentially constant pattern over the year. On the other hand, since the Sun is always low in the sky (if up at all) as seen from Titan's polar seas, there will not be much of a diurnal variation. On that timescale, the tides will have a profound and predictable cycle. Experiments with computer models suggest that solar stirring, wind stress, and tides may all generate separate currents of the order of a couple of centimetres per second, so accurate circulation predictions will have to take all three factors into account.

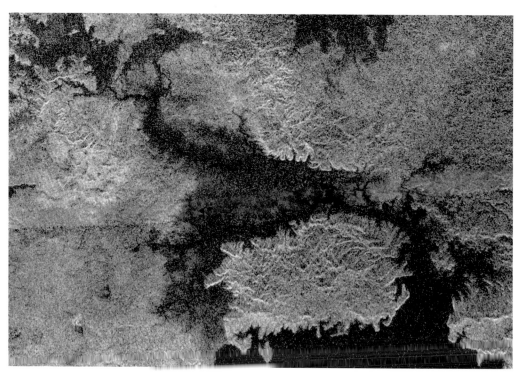

RIGHT Mayda Insula at the northwestern end of Kraken Mare is surrounded on three sides by dark areas that are still radar-brighter than the open sea to the south. The upper channel (Genova Sinus) may be a tidal flat, with a diurnally variable and sometimes negligible covering of liquid. *(Author)*

TIDES

Tides result from the inverse-square nature of gravity. A planetary body's surface (liquid, say) at the point beneath an external body is pulled more strongly than is the centre (and the rest of the solid structure that is mechanically attached). The presence of the Sun and Moon therefore raises tides in the seas of Earth. Similarly, albeit less intuitively, the solid Earth is pulled more strongly by the Moon than the sea at the point opposite. Thus the vector difference in the two accelerations (i.e. the tendency experienced by the liquid relative to the solid Earth) tries to pile the liquid in a bulge opposite the Moon, as well as in a bulge underneath the Moon.

In the simplest idealisation, the horizontal component of the gravity difference is balanced by the height variation of the liquid, and mathematically the liquid will seek to establish what is known as an 'equilibrium ellipsoid'. For the Earth today, the Moon's gravity would make the height of this ellipsoid about 18cm relative to a sphere.

The situation is complicated by a number of factors. Firstly, the Earth rotates 'underneath' this ellipsoid, so a given spot on the planet typically sees two high tides per day. Furthermore, the tides change in time because the Moon (and the tidal bulge that it defines) marches around the Earth once every month. Second and more importantly, the geographical configuration of the ocean basins means the liquid cannot respond perfectly to the changing potential – it is always trying to catch up with its gravitational ideal, causing it sometimes to slosh around. These tidal resonances can selectively amplify different modes or periods, such that some places see only a diurnal tide (one high tide per day) rather than the semidiurnal tide that the two bulges cause. A third, incidental complication for the Earth today, is that the Sun's gravity also causes a tidal bulge, somewhat smaller than that due to the Moon. The overall effect is greater when the two bulges align as a Spring tide.

Earlier in its history, the Earth rotated more rapidly than it does now and the Moon was closer, so the lunar tides were much stronger. The effect of drag on tidal currents has been to drive the Earth–Moon system into a lower energy state. This exported angular momentum from Earth to the Moon, causing it to recede from us (a migration that is not only detectable in the timing of eclipses, but has also been measured directly by laser pulses

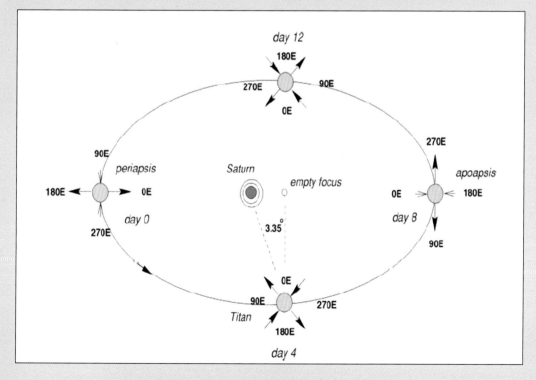

LEFT As Titan rotates synchronously, most of the tidal bulge is almost fixed relative to the surface. However, because Titan's orbit around Saturn is eccentric, the tidal force grows and shrinks, causing the body of Titan to deform and surface liquids to move. (T. Tokano)

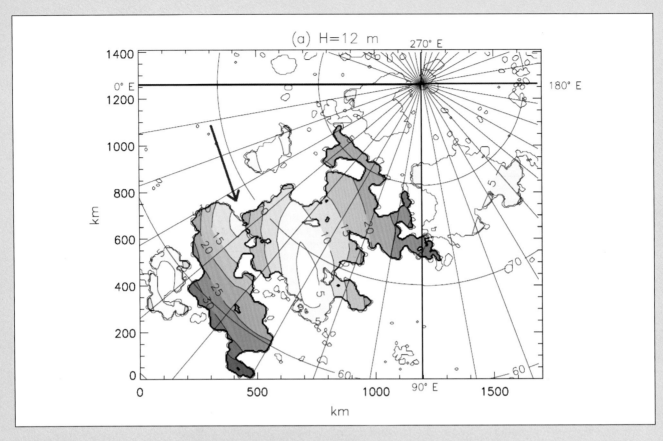

ABOVE The tidal range in Titan's seas (in centimetres) predicted by a numerical model. An important parameter (H) is the assumed depth of Seldon Fretum (arrowed) which controls the flow between the two main basins of Kraken Mare. Being smaller and nearer the pole, the tidal range of Ligeia and Punga Maria and Jingpo Lacus is rather small. *(T. Tokano)*

bounced from reflectors installed on the lunar surface by Apollo astronauts).

Whereas Earth's ocean tides essentially result from the Earth rotating underneath two sets of tidal bulges caused by the Moon and Sun, the situation on Titan is quasi-static. Solar tides are negligible, but the powerful gravity of Saturn leads to a bulge of about 100m in equipotential height at the points directly beneath and opposite Saturn. The fact that Titan's axial rotation is locked to its orbital period means that it maintains the same face toward its primary, therefore this bulge is essentially fixed in position and we wouldn't notice any variations. However, its orbit around Saturn is appreciably eccentric (e=0.029). As the tidal acceleration varies by three times the eccentricity, or about 9%, if there were a global ocean on Titan the tidal bulge would grow and weaken once per orbit, a cycle of 15.945 Earth days. This tidal forcing period is large enough that even relatively shallow basins have natural 'sloshing' periods that are too short to be resonant with the forcing. A tide can be amplified where the basin is sufficiently wide and shallow for this period to be similar to that of the driving tides. This happens in the North Sea, and famously in the Bay of Fundy in Canada, where tides reach ~10m. But it does not happen on Titan.

At the present epoch at least, Titan's seas are of quite limited geographical extent, so the tidal amplitude is not nearly so large as the 10m that a global ocean would see. Ocean circulation models developed for Earth have been adapted to Titan and driven with the appropriate cycle and distribution of tidal accelerations. If Titan's ice crust were completely rigid, then Ontario Lacus would have a tidal range of ~0.4m, and sprawling Kraken Mare might have a range of ~5m. However, because Titan's internal water ocean and the ice crust above it deform as well, the effective change in liquid height on the surface will be reduced by a factor of perhaps 6 or so.

In general, tidal currents are of the order of 1–2cm/s, but they are amplified significantly in the narrow passages of Trevize Fretum and Seldon Fretum to perhaps 50cm/s at peak flood. Some circular currents or gyres will likely occur in Ligeia and Ontario.

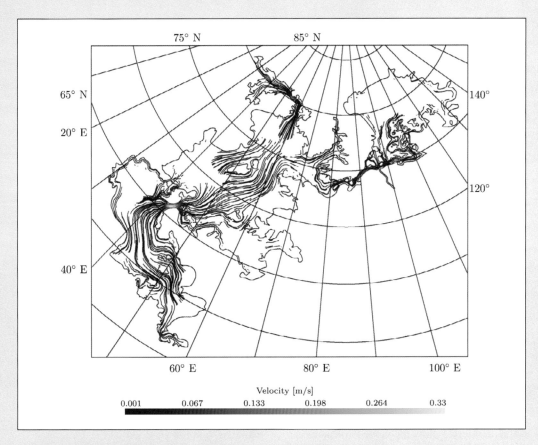

LEFT Simulated streamlines of tidal current at the periapsis in Titan's orbit. Liquid throughout the seas is being drawn to the southwest. A current is not shown when it is less than 1mm/s – generally they are a few cm/s, except where the flow is constricted by Seldon Fretum, initially referred to informally as 'The Throat of Kraken'. *(D. Vincent)*

BELOW Although it is a perfectly calm day, the sea in the Corryvreckan Strait between the Scottish islands of Jura and Scarba is rough with breaking waves and whirlpools resulting from the energetic tidal flow through the channel. The tidal flow through Seldon and Trevize Freta may cause similar rough seas. *(Author)*

An important determinant of the tidal dynamics of Kraken is the width and depth of Seldon Fretum. If it is a shallow, constricted channel, the two large basins are essentially isolated and will behave as separate bodies. On the other hand, if it is more than a few tens of metres deep, liquid will be able to surge through and the depth profile across Kraken will see-saw with the tidal potential. In the case of Trevize Fretum, it is probably sufficiently long and constricted to essentially decouple Ligeia and Kraken, although at least some liquid will surge through.

It is interesting to speculate how the dynamics might have changed over the past millennia. If the sea levels in Kraken and Ligeia were lower by ~10% (say), Trevize Fretum would have been high and dry and the two seas would have been completely separate. Conceivably some future colonists of Titan could exploit the tidal power available in Seldon Fretum, but most likely they would be reliant on nuclear power anyway for other reasons.

Astronomically, the dissipation of energy in Titan's seas (and, indeed, its interior) should lead to a reduction in the eccentricity of Titan's orbit. Indeed, the fact that it is not already more circular than it is has prompted speculation that it is being maintained by a barely perceptible resonance with Jupiter, or was nudged up by Titan undergoing a recent close encounter or a giant impact.

RIGHT Experiments by this author in the Mars Wind Tunnel at NASA's Ames Research Center in 2003 showed that capillary waves are generated more readily in liquid hydrocarbons than in water. The apparatus (on the left) has small sensors to measure wave height positioned above a tray of kerosene. The wind is flowing left to right.
(Author)

BELOW Wind predictions (m/s) for the north of Titan show relatively low values during the nominal mission, but rising possibly to levels where waves should form in northern summer.
(Author)

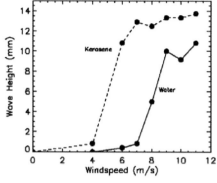

Waves

The possibility of waves on Titan's seas was recognised during the formulation of the Cassini mission, and so the Huygens probe was equipped with tilt sensors to measure any motion on waves in the event of it surviving splashdown on a liquid surface.

The study of ocean waves on Earth had a rather empirical background because, until a probe was designed for Titan, there had been no need to consider waves in liquids other than water, driven by gases denser than sea-level air, in gravity other than Earth's. Historically, before the development of meteorological instrumentation, winds at sea were estimated by the height of waves using the Beaufort scale.

On Titan, everything is different because in the low gravity the propagation of a wave of given wavelength is rather slow compared with Earth, much as a pendulum will swing slower there. Similarly, a wave of a given energy will be higher in amplitude, and so waves of a given height can be generated by lower winds on Titan than are required on Earth. However, other factors may be important too. For example, the possible role of atmospheric density in wave formation and growth was explored in some wind tunnel experiments prior to Cassini's arrival.

Observations of the Sun's reflection on the sea surface, and radar observations, have indicated that Titan's lakes and seas are generally as flat as a millpond. Indeed, one iconic image in the near-infrared of the Sun glinting off a flat liquid surface earned that particular lake the name Jingpo Lacus, after a Chinese 'Mirror Lake'. If the hydrocarbon liquids behave somewhat like water in the presence of the low gravity and thick atmosphere, the absence of waves in Titan's seas posed a puzzle.

Actually some transient roughening has been observed here and there, most likely due to the formation of 'catspaw' or incipient capillary waves. Detailed studies found that the threshold wind speed for wave generation ought to be ~0.4m/s for methane-rich (low viscosity) seas, or ~0.6m/s for ethane-rich seas (a viscosity similar to water). Such speeds probably only occur in the northern hemisphere near its summer solstice. At that time in 2017 it was hoped that more robust evidence of waves would be observed by Cassini over Kraken or Ligeia toward the end of its mission, but we appear to have been unlucky. It may be that the wind freshening occurs later than some models predicted.

The places where roughening was seen in the near-infrared by VIMS are in Punga Mare, some areas of Kraken Mare, and in and near Trevize Fretum. The latter saw roughening of a plume, apparently flowing into Ligeia. Roughening indicated by radar was seen quite early in the Cassini mission in the Moray Sinus estuary at the north end of Kraken, and on a few occasions just off the jagged coast on the south side of Ligeia Mare. The spot where brightening came and went was nicknamed the 'Magic Island'. Winds seem to be the most likely explanation for this roughening of the surface, but currents driven remotely by wind stress elsewhere could do it. More exotic scenarios such as bubbling or pumice rafts cannot be ruled out. What is known, is that while some subsurface scenarios could be considered for the radar bright spots, at least in those spots where bright off-specular glints were present in the near-infrared, there must be roughness at the surface. Of course, the simplest solution would be for both near-infrared and radar indications to have the same explanation, and the more numerous detections later in the Cassini mission argued, at least circumstantially, for a seasonally related origin. Although this could, in principle, include thermal 'stirring' of the seas by sunlight, the windier conditions of summer were more likely responsible.

The force that pulls the wind-stressed surface back to horizontal in the first, smallest waves is surface tension. Capillary waves having wavelengths of a few centimetres can be particularly important on both Earth and Titan in controlling radar reflectivity, since a corrugated surface can act like a diffraction grating (Bragg scattering). Once capillaries form and grow, they can start to push energy into longer wavelength waves, where gravity is the restoring force; these are the more familiar chop and swell in the sea. Wave-wave interaction and other dissipative processes will limit wave growth, but the principal effect that controls wave height in a given wind is that a growing wave will run faster and cause the relative speed of the wind over it to

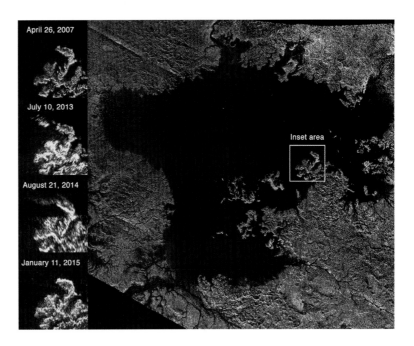

ABOVE The 'Magic Island' in Ligeia Mare is a feature located at the tip of a promontory that changed in appearance much more than the different viewing conditions would lead us to expect, in turn suggesting a change in the sea surface, perhaps due to waves formed by wind or currents veering around the promontory. The box is 60km across. *(NASA/JPL-Caltech/ASI/Cornell)*

BELOW Specular reflection from Titan's surface in VIMS data (red corresponds to 5μm light, which most easily passes through the haze). The bright concentrated reflection of the Sun indicates a flat sea at this point, but a few dimmer points are seen toward 11 o'clock, indicating glistening (rougher) solid surfaces. *(U. Idaho)*

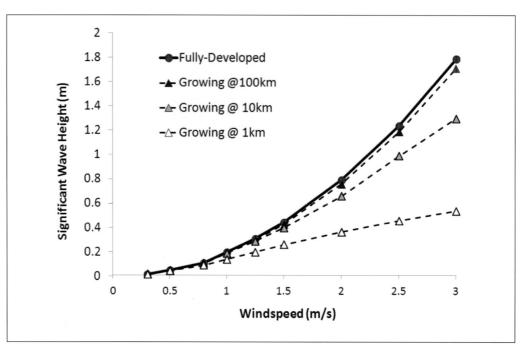

RIGHT On Titan, winds of less than ~0.5m/s will not produce waves at all, but stronger winds will yield waves larger than those which occur for the same speed on Earth. The curves show wave height after fetches of 1, 10 and 100km – in Titan's dense atmosphere, waves grow almost to their fully developed height after 10km or so.
(Author)

decline. Consequently, growth becomes self-limiting in a so-called 'fully developed' sea. This state should be reached after a fetch of the order of 20km on Titan, so in small lakes and at the margins of the larger seas the waves are still growing.

There ought to be a fully developed wave spectrum spanning most of the interiors of Punga, Ligeia and Kraken. The fully developed 'significant wave height' (SWH, a kind of weighted average) should be ~20cm for a 1m/s wind. The strongest large-scale wind predictions at the surface from circulation models are ~2m/s, giving a wave height of 80cm. On rare occasions localised methane storms could give stronger winds.

From classical physics, we know that the superposition of lots of random waves gives rise to a statistical distribution of surface heights (the Rayleigh distribution). By this, about one out of 1,000 waves may have a height that is twice the SWH, and occasionally even larger 'rogue' waves will be encountered. These must be taken into account when estimating coastal erosion on Titan – and indeed when designing vessels that are to sail its seas.

As on Earth, liquids on Titan should facilitate the motion of solid sediments, both by dragging particles and, via buoyancy, by reducing their effective weight. The terminal velocity of ice or organic particles in hydrocarbon liquids will be rather low, due to the modest density contrast and the low gravity. As a result, the fluid currents required to move sediments are quite small. On the other hand, liquid of a given depth flowing along a given slope will be slower on Titan than on Earth. Hence, we should expect sediment mobility and sedimentary structures such as sandbars and beaches on Titan to be much as they are on Earth.

Since the relative humidity of methane on Titan is only ~50%, a body of pure methane cannot persist indefinitely on Titan's surface. Estimating on the basis of terrestrial empirical transfer coefficients it will evaporate at around 1m/year, although this is strongly dependent on wind speed. The evaporation rate is composition-dependent, in that the saturation vapour pressure of ethane is very low, so ethane acts to suppress the partial pressure of methane above mixed-composition seas (much as syrup will evaporate in a kitchen much more slowly than water).

An interesting subtlety of Titan's seas is that nitrogen dissolves appreciably (~30%) in liquid methane, so heat or mixing methane-rich and ethane-rich fluids may cause effervescence. In fact, a 2° change in temperature of methane (such as the waste heat of a vehicle) will drop the solubility of nitrogen by 2–3 grams per litre, or roughly by the amount of CO_2 in a carbonated drink. So a boat for Titan that does not have adequate insulation may cause the liquid to fizz around it, and if the instrument suite of the boat includes a sonar this might interfere with its operation.

Since nitrogen is barely soluble in ethane, one can imagine scenarios where nitrogen-saturated methane (rainfall, coalesced into a flash flood in Vid Flumina, perhaps) flows into an ethane-rich body of liquid. As they mix, the permitted amount of nitrogen drops and so the nitrogen comes out of solution as a froth of bubbles. Desaturation processes do not always occur gently. For example, CO_2-saturated water at the base of Lake Nyos in Cameroon suddenly overturned in 1986, smothering a nearby village in a blanket of gas. At least occasionally, Titan's seas may exhibit similar exotic violence.

The possibility of layering in Titan's seas is an interesting one. Just as our seas aren't perfectly mixed (for example, the dark anoxic waters of the Black Sea) so there could be an ethane-rich layer in Titan's depth – perhaps wily captains of future Titan submarines will exploit such an analogue of the 'thermocline' to elude detection by sonar from ships hunting them!

It isn't hard to imagine that in the last reversal of Titan's climate cycle, 30,000 years ago, when most (or all) of the liquid methane was in the southern hemisphere, some ethane-rich Ontario-like remnants might have occupied the deepest parts of the dry Ligeia and Kraken basins, and maybe as those seas

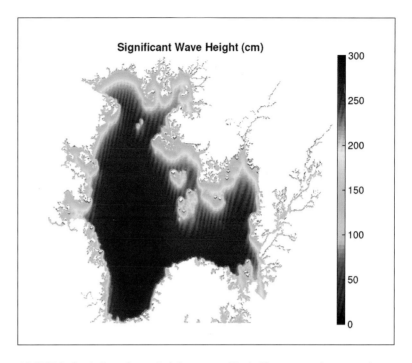

ABOVE A simulation of wave heights across Ligeia Mare, assuming a steady 3m/s wind from the top of the picture. The strongest waves are found at Wakasa Sinus, the westernmost bay, where the waves have had the longest fetch over which to grow. Models like this can be compared with the detailed appearance of the shoreline to determine the relative influence of waves, tides, and other processes. *(Alex Hayes/Cornell)*

started to fill, the ethane remnants were not vigorously mixed and lurked unbuoyant in the depths. These dregs would likely be rich in other solutes, so just as on Earth gypsum, limestone, and even silica rocks dissolve in small amounts in water, so there may be some butane, benzene and nitriles dissolved on Titan.

 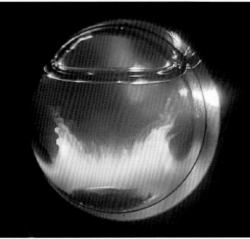

LEFT Although ethane, methane and nitrogen are nearly completely colourless, their interactions can lead to bubbles and clouds of tiny droplets. These experiments were done at 80K, which is colder than Titan, with nitrogen gas being added to a methane/ethane mixture and dissolving. *(Jennifer Hanley)*

Chapter Seven

Restless atmosphere

It is its atmosphere that makes Titan unique. Not only has the atmosphere shaped the landscape we see, but it continues to do so and is always changing. Occasional rainstorms puff up over hours, while seasonal changes in winds and the haze make Titan's appearance change from year to year.

OPPOSITE Equatorial methane clouds seen in 2010 shortly after the northern spring equinox on Titan. This is the Saturn-facing hemisphere of Titan. The crater Sinlap can be seen just above left-of-centre. *(SSI/NASA/JPL)*

RIGHT A schematic of the processes occurring in Titan's surface and troposphere (up to ~40km) and the stratosphere above. Graphics like this began simply in the Voyager era, then became more complex as new processes were recognised. *(APL)*

BELOW In this schematic of the nested cycles on Titan there is a purely physical inner hydrological cycle of methane evaporation and condensation that links to the photochemical destruction of methane and its possible geological replenishment. *(Author)*

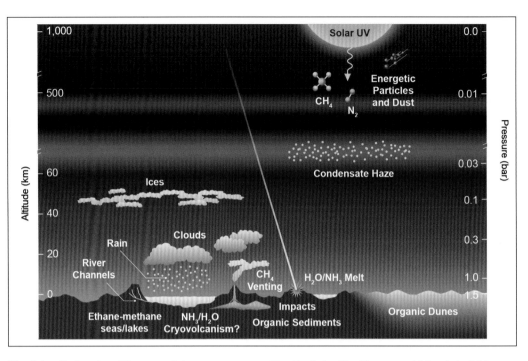

The thing that makes Titan special among planetary moons is its atmosphere, its 'ocean of air'. And remarkably, this bears a striking physical similarity to that of Earth – being just a modest factor denser and being made of the same principal constituent, molecular nitrogen. Not only does Titan have a 'warm' stratosphere like Earth (unlike Venus and Mars), but it has a strong greenhouse effect which is forced by a condensable vapour. And the abundance of this vapour is sufficient to create clouds that don't just trace atmospheric motion, but carry enough energy to force it. Such vigorous moist convection occurs on the giant planets, but only on Titan and Earth is there precipitation falling onto the surface. The next time you grimace at a rainstorm, consider it a privilege to be domicile on one of only two worlds in the solar system where this happens!

Greenhouse structure

The overall temperature structure of an atmosphere depends on how transparent (or otherwise) the gases and particulates are to sunlight and to thermal (infrared) radiation. The absorption of sunlight by haze makes the stratosphere on Titan relatively warm, and the thermal opacity of the lower atmosphere (i.e. the greenhouse effect) blankets in the sunlight that is absorbed there by the surface and by methane. In between, there is a temperature minimum at 40km altitude, the tropopause, at ~70K. This 'cold trap' profoundly influences the abundances of gases. For example, organic-rich air welling down from where the compounds are created by photolysis altitude of 100km, and if it reaches saturation it will form thin layers of cloud.

Similarly, methane that is welling up from the surface can condense upon encountering cooler temperatures. By limiting the methane abundance to ~1% the tropopause throttles the delivery of methane to the stratosphere, and destruction by photolysis. Thus the temperature structure controls the distribution of gases, just as the gases control the temperature structure. Similar coupled feedbacks influence Titan's circulation (i.e. winds) and particulates (haze and cloud). All these interactions make Titan such a complex world that is challenging, but instructive, to model.

Just as the major contributor to Earth's greenhouse effect is the working fluid for our weather, namely water (vapour), so the major factor in Titan's greenhouse effect is methane. In fact, the low temperatures on Titan mean that thermal radiation peaks at quite long wavelengths, and the blanketing effect of the main gas (nitrogen) is significant too. Hydrogen, even though it is present only in relatively small amounts, also contributes some opacity (in combination with nitrogen) at shorter wavelengths. The reason methane is so important is that the wavelengths which it absorbs most effectively lie in the gap between the hydrogen-nitrogen and nitrogen-nitrogen 'windows'.

In the lower atmosphere of Titan where the thermal opacity prevents the ready escape of heat by radiation, the energy deposited on the surface by sunlight is transported away by another process, namely convection.

It is convection that is the engine of the weather system on Titan. Warm air rising draws air from nearby to replace it, making surface winds. And rising moist air can drive storms which cause locally stronger winds, as well as rain. Broadly speaking, scientists underestimated the vigour of Titan's weather, in part because the early models of the energy balance considered only global-average, annual-average fluxes and suggested only 1% of the top-of-atmosphere sunlight would drive convection. When that heat of ~0.04 W/m^2 was expressed as the latent heat of evaporation of methane, it corresponded to a mere 1cm/year of rainfall; as compared with

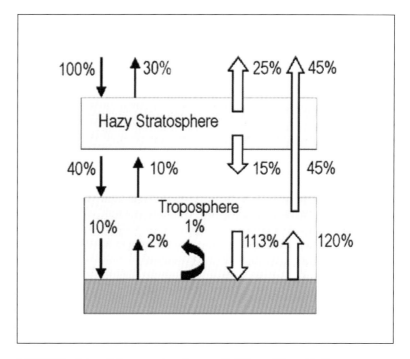

ABOVE The thin solid arrows show how sunlight is partitioned in the atmosphere of Titan by scattering and absorption (expressed as a percentage of the flux incident on the top). White arrows show how thermal infrared (plus the black curled convection) balances out this energy flux. *(Author)*

BELOW A near-infrared view of Titan in September 2010, shortly after spring equinox, shows a bizarre arrow-shaped storm. *(NASA/JPL/SSI)*

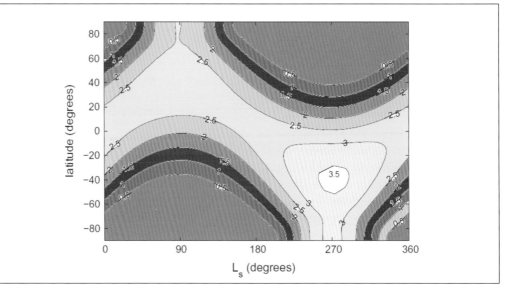

RIGHT Sunlight at Titan's surface against season and latitude. Although the sunlight at the top of the atmosphere actually peaks at the pole, the scattering in Titan's haze means the flux reaching the surface is a maximum at midlatitudes. The pattern shown here, with the overall maximum in the south, was inverted about 30,000 years ago. *(Author, from calculations by J. Lora)*

the ~100W/m² of convection on Earth that amounts to about 1m/year of rain.

But the 'difference of averages' is not the same as the 'average difference', because despite the small fluxes involved, there is strong temporal variation. The thermal fluxes change very little during the course of a Titan year because the dense thick atmosphere responds only slowly to changes, but the sunlight changes dramatically with season. As a result during polar summer there is a huge imbalance where the strong absorption at the surface (~1W/m²) of sunlight isn't offset by thermal fluxes, and so must be balanced by convection. Hence clouds and rain follow the strongest sunlight, which occurs at the respective pole in the summer and at the equator in spring.

Cloud formation and rainfall

When a parcel of warm moist air rises, it cools because it expands into the lower pressure. The ability of the moisture to remain a gas depends on the saturation curve of the fluid, which is a steep function of temperature. For example, on Earth water has a saturation vapour pressure of 6 millibars at 0°C, but 20 millibars at 20°C. So if we have air at sea level (~1,000 millibars) with 50% relative humidity, then 1% of the air is water vapour (0.5×20/1,000). Now, if we raise this parcel of air by a couple of kilometres, say by winds pushing it over a mountain range, it will drop in pressure to ~800 millibars and cool by ~20°C, with the result that now the 1% of water

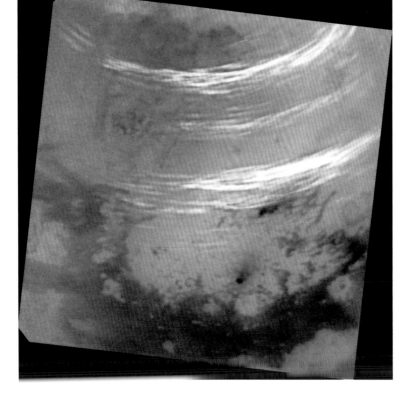

LEFT A near-infrared (938nm) image of Titan taken by Cassini in 2017 from a range of 508,000km. Centred at 57°N, it shows streaky clouds in the northern hemisphere, where it was late spring. The north pole and seas are visible at the top. *(NASA/JPL/SSI)*

vapour corresponds to 8 millibars, or a relative humidity of 133%, prompting the condensation of droplets. This is why so-called 'orographic' clouds form on the lee side of mountains.

Condensation can also occur when air rises after the surface has been heated by sunlight. Of course, by the logic above, warm conditions have a higher saturation vapour pressure, so it takes more actual fluid material (i.e. a higher percentage or absolute humidity) to reach 100% relative humidity, and thus the condensation point. So cloud formation is a delicate balance between the amount of moisture and the temperature.

However, once it happens things can get rather interesting. Because the condensation process releases latent heat (the energy required to evaporate the fluid in the first place), this heat can sustain the convective process. Put another way, moist air that is starting to condense will not fall in temperature as fast as dry air would, so the moist air remains buoyant and continues to ascend, causing more cooling and, if there is enough moisture, more condensation. That is, the process feeds back on itself, forming towering cumulonimbus clouds that can push all the way through the troposphere. All this physics for water on Earth is exactly the same as for methane on Titan, except of course all the numbers are different!

The individual condensation drops are initially tiny. There is an activation energy required to cause the molecules to arrange themselves into a drop. It can be lowered by having something to condense onto, such as a dust grain, a salt particle, or even an ion produced by a cosmic ray. The physics will depend upon very specific material properties, and so could be very different on Titan than on Earth – and thus difficult to predict. But some experiments that attempted to condense methane onto tholin (organic) particles showed that things aren't so much different. While condensation will occur on Earth quite robustly at humidities of even 101%, the energy barrier is only slightly larger for Titan, and perhaps permits saturations of 130% before liquid starts to form.

Once some drops have formed, it is easier for vapour to condense onto those than to start tiny new ones afresh. But as soon as you have a few drops a little bigger than the others, they fall a bit faster relative to the air and the difference in fall speeds (whether they are in a parcel of air that is rising relative to the ground or not) makes collisions possible. Colliding cloud droplets will likely stick together, and grow first into drizzle and eventually raindrops. Of course, this situation becomes rather more complicated when local conditions are near the freezing point, because it will permit hailstones much bigger than raindrops to form, but again the process is likely to be similar on Titan.

The kinetic energy of air is effectively the output of a steam engine driven by humidity. The updraught within a cloud may be about 10m/s on Earth, and perhaps a factor of two weaker on Titan. This modest difference is rather remarkable, given that about 1,000 times less sunlight reaches the lower atmosphere of Titan than of Earth. The physics of convection are such that this weak forcing power is manifested on Titan in a much smaller area fraction of convective plumes than on Earth, making clouds and rain on Titan correspondingly much rarer than on Earth. There are large seasonal and latitude variations on Titan, but it might be fair to say the cloud cover is of the order of 1% or less on average, compared with about 30% on Earth.

As long as the fall velocity of a raindrop (or hailstone) is less than the updraught, it can remain in the cloud and grow. Some will randomly fall out of the bottom (or sides) of the cloud, and of course the updraught may die away and release the drop population. The maximum size of the drops is limited by a balance between surface tension of the liquid that holds a drop together, and the aerodynamic drag forces that tend to tear it apart. For very small droplets which fall slowly, the surface tension dominates and the particles are perfectly spherical. However, the larger drops fall faster, leading to greater drag. As this acts substantially on the bottom of the drop, a drop in the sky is a somewhat flattened shape – although ironically it isn't a 'teardrop' shape! Eventually, the drag pressure on the bottom 'blows out' the centre, causing the drop to break apart.

There is not a hard cut-off, of course. The breakup is a statistical process depending a

Stephen Baxter's novel *Titan* features a mission to Titan, and describes astronauts paddling around a methane-ethane lake in a boat improvised from a heat shield. Published in 1997, it predated the Cassini-Huygens results. Drawing insight from a paper by the present author, Baxter made one of his characters provide a vivid description of rainfall:

"The biggest drops were blobs of liquid a half-inch across. They came down surrounded by a mist of much smaller drops. The drops fell slowly, perhaps five or six feet a second... visibly deformed into flat hockey-puck shapes, flattened out, she supposed, by air resistance..."

BELOW The overall tropospheric meridional circulation has a cross-equatorial flow for most of Titan's year, with upwelling over the summer pole. A brief interval of more symmetric circulation (like the Hadley flow pattern on the faster-rotating Earth) occurs for a year or two near the time of an equinox. *(Author)*

2003 - Southern Summer

Single pole-pole Hadley cell

Downwelling of organic-rich air over northern stratosphere; high, layered clouds (ethane?)

Cross-equatorial surface winds

Upwelling, convective clouds and rainfall over south

2009 - Northern Spring

Quasi-symmetric Hadley circulation around equinox season

Equatorial rainstorms and dune-sculpting winds

little on the history of swirling inside the drop. It is possible to get the occasional 'superdrop' outlier, but basically the balance is defined by the Bond Number. This leads to a limit of about 6.5mm diameter for water on Earth falling at about 10m/s. The surface tension of methane on Titan is weaker than water and the low gravity means the drag is weaker still, enabling drops to grow to 10mm diameter and to fall at 1.5m/s, which is about the same speed as a snowflake falls on Earth.

Simulations show that cloud systems may develop over timescales of several hours, rising to heights of a few tens of kilometres. Because the drops fall so slowly, they take some hours to reach the ground – so long that a cloud system can develop and collapse before the first drops land. The instantaneous rain rates can be rather high, depositing tens of centimetres of rain in an hour or two. An individual rain shaft can be a few tens of kilometres across.

The principal source of lift on Titan seems to be solar forcing, with the cloud pattern broadly following the seasons. The latitude of the subsolar point, where the strongest instantaneous heating occurs, swings from 26°N to 26°S. The strongest daily-averaged heating would be at the summer pole (as on Earth), although the thick, hazy atmosphere means that the strongest surface heating does not quite reach that far poleward. In any case, strong cloud activity was observed around the south pole during southern summer – first in the period 2000 to 2005 by adaptive optics telescopes, and in late 2004 by Cassini as it arrived.

Rising air is the typical prerequisite for rain (we shall return to some other cloud types in the stratosphere associated with descending air in the following chapter). Mountains may trigger clouds and precipitation, as occurs on Earth. Another scenario is where moisture is added to the air from a surface source – a prominent illustration of this being the 'lake-effect' snow seen in the northwestern USA. Some observations have hinted at this possibility on Titan associated with the northern seas. Prior to Cassini, there were even some suggestions of preferential areas for clouds at low latitudes, perhaps indicating a localised source of

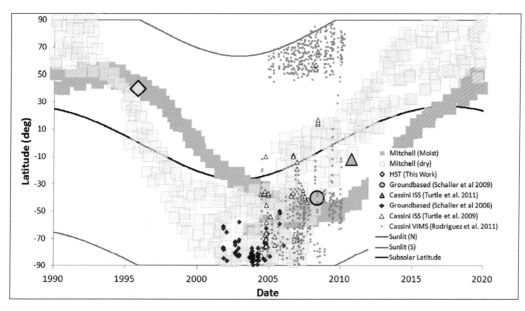

LEFT Cloud and precipitation patterns over a Titan year. The areas shaded blue and pink are global circulation model predictions for a moist and a dry atmosphere, respectively. The observations appear to support an overall rather 'dry' picture, but more advanced models suggest the high latitudes may be at least somewhat 'moist'. *(Author)*

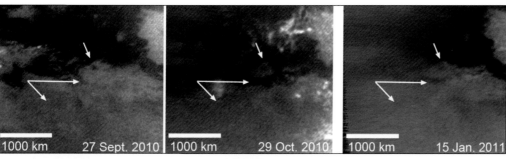

LEFT April showers on Titan: in this sequence shortly after equinox, surface darkening and re-brightening was seen in Concordia Regio near where cloud activity had also been observed. It is believed an area approaching 500,000km² was temporarily darkened by being flooded. *(NASA/JPL/SSI/U. Arizona/E. Turtle)*

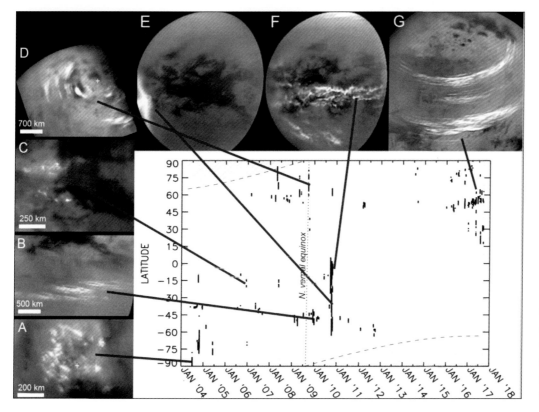

LEFT The different styles of cloud observed with the Cassini ISS have been mapped onto the latitude and season at which they were observed. The cloud morphologies are quite different at different times. *(E. Turtle)*

RESTLESS ATMOSPHERE

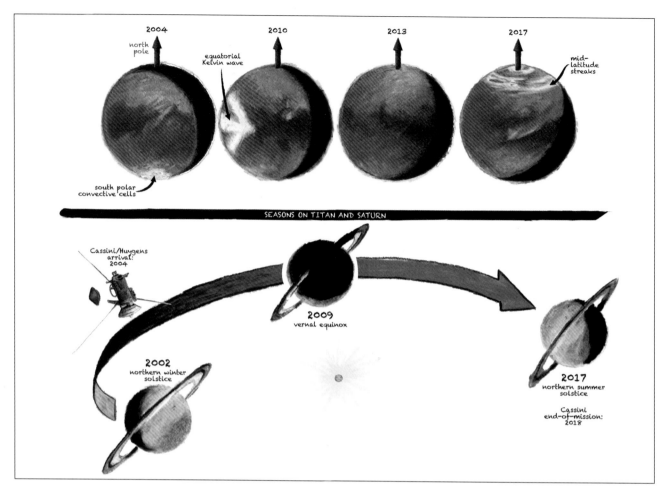

ABOVE A summary of the different styles of methane cloud activity during different seasons on Titan and Saturn.
(James Tuttle Keane)

methane, but the possibility of wetlands or geysers has remained speculative.

The south polar activity coyly shut down soon after Cassini's arrival, but diligent monitoring showed several large cloud events at the equator in 2009. That being the equinox, the subsolar point and the generally rising air known as the 'intertropical convergence zone' were both on the equator.

> The most impressive meteorological phenomenon was the so-called 'Methane Monsoon', which often – though not invariably – occurred with the onset of spring in the northern hemisphere. During the long winter, some of the methane in the atmosphere condensed in local cold spots and formed shallow lakes, up to a thousand kilometres square but seldom more than a few metres deep…
>
> **Arthur C. Clarke** *Imperial Earth*, 1975

One such event seemed to darken a streak of the surface thousands of kilometres in length that lasted for several months. This was interpreted as a temporary darkening of the ground by the dampness (and perhaps puddles) of a rainstorm.

One final trigger for the onset of rain can be dynamic – waves propagating in the atmosphere (a prominent example at mid-latitudes on Earth is the wiggling jetstream that is associated with a wave period of a few days). A particular kind of wave has been suggested as being responsible for a giant cloud system observed on Titan in 2010 that had an unusual arrow shape.

The most recent climate models suggest that the character of rainfall may vary systematically with latitude. This prediction seems to be consistent with the detection (only at mid-latitudes) of alluvial fans, which are landforms that require abruptly enhanced levels of streamflow and thus transient rainfall. The rainfall at the poles seems to be rather more steady.

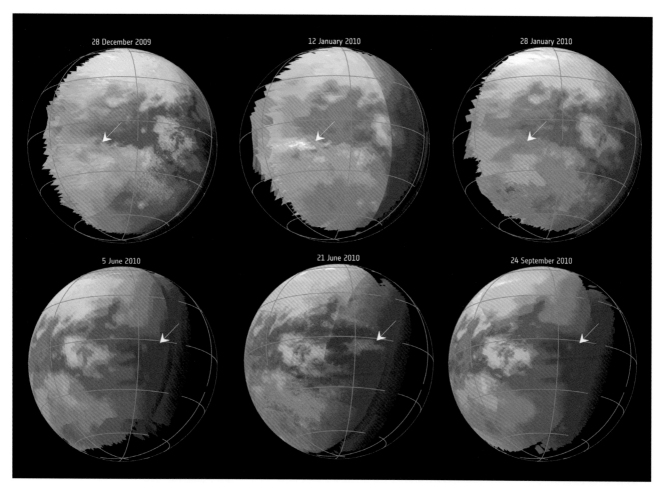

ABOVE **Two instances of near-equatorial storms seen close to equinox, depicted in false-colour sequences of VIMS data. In this case, red highlights the opacity in the very lowest part of the atmosphere, interpreted as a haboob.** *(NASA/JPL-Caltech/U. Arizona/U. Paris Diderot/IPGP/S. Rodriguez)*

When a raindrop falls at a steady speed, its weight is balanced by drag, making it continuously dump downward momentum into the air around it. Any evaporation of the drop in dry air can cause cooling and thus a downwelling, in a reversal of the process that formed the cloud drops in the first place. Thus, especially in the late stages of a storm, there can be a strong 'downburst' of air from the base of the cloud. When this hits the ground, it has nowhere to go but outward, forming a layer which flows radially away from the rain core. In regions where there is surface dust, this outflow can pick up dust to form a dust storm or 'haboob'. One Cassini observation indicated a localised obscuration low in the atmosphere that may have been such a dust storm.

Although individual storms may be rather localised, their aggregate effect helps to shape the most prominent of Titan's landforms, namely the equatorial dunes. A major puzzle early in the Cassini mission was that the dunes gave the impression that the sand transport was invariably eastward, with the regional deviations of 10–30° presumably being caused by local topography. The linear form implied bidirectional winds at the surface, consistent with a seasonally shifting north–south pattern, but at low latitude and low altitude the pattern would be expected to have a slight westward bias (akin to the trade winds on Earth). The answer to the paradox is that the dunes only reflect the strongest winds – as on Earth, the sand only moves in the windiest few per cent (or less) of the time. The strongest winds that a given spot on Titan's equator is likely to see are those associated with a storm. These will be (effectively randomly) locally outward, but will also carry some of the eastward momentum in the atmosphere more generally, hence the aggregate effect is eastward. Although the winds measured by Huygens during its descent were slightly westward in the lowest 7km, the zonal winds measured aloft were more strongly eastward, as expected. In effect, a rainstorm (or

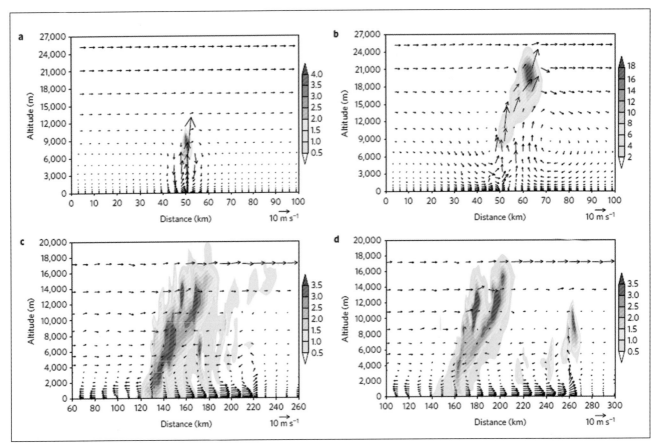

ABOVE A high resolution simulation of an equatorial rainstorm. The colours indicate the amount (g/kg) of methane condensed as cloud, hail and rain snapshots at (a) 1hr, (b) 1hr 40min, (c) 10hr 35min, and (d) 12hr 30min after the start of the simulation. The initial updraught forms a big cumulus tower cloud, then rain falls. A downdraught forms, dragging eastward momentum down from the high altitudes to form a strong jet hugging the ground that may form a haboob, and gives the dunes their eastward trend. *(B. Charnay)*

convective activity at equinox more generally) 'mixes downward' some of this eastward flow and overcomes the general westward tendency.

Temperature variations and winds

The air temperatures at Titan's surface are relatively uniform at the 94K measured by Voyager and by Huygens, in part because the thick, dense atmosphere holds a lot of heat and requires a long time to warm up or cool down. Nevertheless, the surface itself does respond a little to the direct heating by the Sun, and both models and observations indicate that polar summer may reach as high as 97K and the polar winter perhaps as low as 91K. Although pure methane and pure ethane will both freeze at ~90K, a mixture of them, or a mixture of methane and nitrogen, will require a lower temperature, so the polar winters are not cold enough to induce freezing out.

At the equator of Titan, as on Earth, there is relatively little seasonal change in temperature. Between day and night, there is a hint in Cassini's infrared data of a ~1K variation in surface temperature in the large dunefields. These would be expected to have the biggest change for two reasons. Firstly, they are dark and therefore able to efficiently absorb sunlight. Secondly, because the sand is porous and likely composed of organic material, it will have a lower heat capacity than solid materials, and lower than water-ice in particular.

The other factor that influences ground temperature is moisture. If the ground is damp, then absorbed sunlight may get soaked up by evaporating the moisture, rather than by raising the temperature. When working in 'listen-only' mode as a microwave radiometer, Cassini's radar could measure surface and near-subsurface temperatures. The relationship of temperature to latitude neatly agreed with circulation models which seem to need a moist surface (at least at higher latitudes) in order to produce the observed patterns of clouds.

The windspeeds in a dense atmosphere driven by only weak solar fluxes were expected to be small – at the surface winds are usually rather less than 1m/s. The indications from

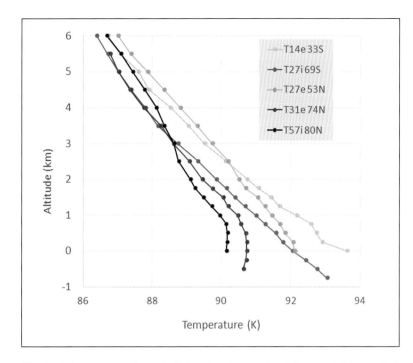

LEFT Temperature profiles in the near-surface atmosphere measured by Cassini's radio occultation experiment (altitudes relative to a 2,575km sphere; Titan's shape means the poleward profiles reach below zero). The warmest profile was at low latitudes, whereas the coldest was in the north polar winter. *(Author)*

BELOW Wind probabilities calculated by three different weather models for Ligeia Mare in 2023. Although the details differ, the overall predictions converged nicely once the winds at different model layer altitudes were corrected to a 10m reference height. The blue dashed line is the conservative analytic model based on these results that was employed in planning the Titan Mare Explorer (TiME) mission. The winds exceed 0.8m/s only 5% of the time. *(Author)*

BELOW Mean near-surface winds for Titan from a circulation model. In the high summer latitudes the winds are less organised, but there is summerward flow over the rest of Titan. As a result, at the equator, the mean winds seasonally alternate between southward and northward, with a slight westward inclination (the opposite sense to that of dune growth). *(T. Tokano)*

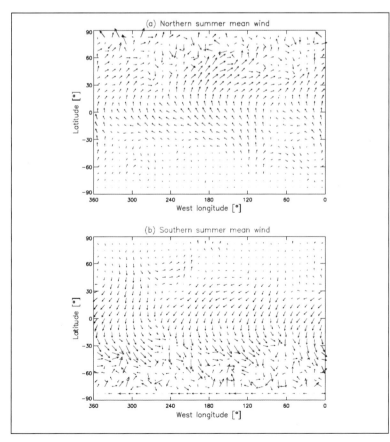

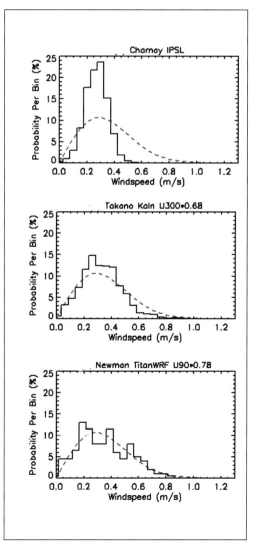

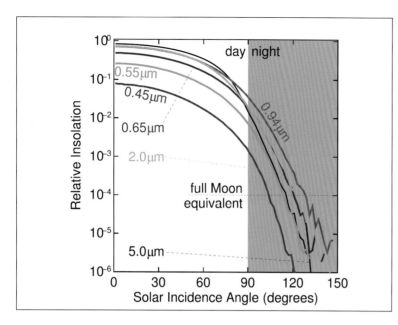

ABOVE Simulations tracking the scattering of light in Titan's spherical atmosphere. An appreciable amount of light gets around to the nightside. At 120° (i.e. a local solar time of 8pm or 4am at the equator) there is more illumination at red wavelengths than Earth sees from full moonlight. *(Jason Barnes)*

Huygens were of winds about 0.5m/s several hundred metres above ground level and maybe 0.3m/s at the surface itself. Theoretical models and wind tunnel data imply that winds of about 1–2m/s are required in order to blow sand around, so the abundant dunes tell us that the winds must occasionally get that fast. In the outwash from a methane rainstorm, a location on Titan might see winds of ~5m/s, but only for a few hours in a Titan year (one-ten-thousandth of that time).

Although the dominant wind in the atmosphere of Titan above ~10km is the prograde zonal wind (eastward), the winds at the surface may be predominantly north–south because of the seasonally forced meridional circulation pattern. This alternating flow, biased with eastward momentum and downward mixing in the equinox storms, gives rise to the eastward-oriented linear dunes.

Of course, as on Earth, there will be weather systems and waves that cause variation from the mean pattern, and the topographical features will cause local deviations that funnel the wind between obstacles. Even over flat terrain, the variations in albedo or thermal inertia may cause diurnal wind patterns – for example, since the dark, porous dunefields heat up during the day by more than their surroundings, this may cause a sort of inverse 'sea breeze' circulation, with an updraught over the dunes in the afternoon drawing in air from round about.

Twilight and visibility

It is important to recognise that Titan's atmosphere is somewhat opaque not because it is very hazy, but because it is thick. Looking in from space, we peer through more than 150km of air, and it doesn't take much suspended material per cubic metre to produce a substantial opacity over such a distance.

To put it another way, the surface contrasts observed by the Huygens probe during its descent were appreciably diminished only above about 10km altitude; that is, the 'visibility' was about this distance. On Earth, this visibility would not qualify as being 'hazy' (2–5km), and certainly not 'foggy' (<1km).

Overall, because Titan's atmosphere is so thick, it has an optical depth of 5 in reddish light (~700nm) and 9 at blue/green wavelengths, rather like a heavy overcast on Earth. In the near-infrared, at 940nm where Hubble and Cassini made their maps, the optical depth is ~3. At the 5μm wavelength of VIMS, it is only ~0.2. By comparison, the Martian sky has a typical optical depth of about 0.7 in visible light, but it increases significantly during a dust storm.

Using the naked eye, you would certainly be able to tell where the Sun was located in the sky, but even with night-vision goggles at 940nm, only a fraction of 1% of the available photons are not scattered by the haze. So, despite the desire of space artists to portray Saturn in Titan's sky, the view would be too blurred to see an 'edge' of the disc. Furthermore, depictions of the rings at a tilt are inappropriate because Titan's orbit is essentially in the same plane as the ring. And even if they were visible, the rings would be perpendicular to the horizon when viewed on the equator and horizontal when viewed at the polar regions.

The fact that Titan's axial rotation is synchronised with its orbital period means that Saturn is almost fixed in the sky – the only motion being a rocking east and west by about 3° because of the orbital eccentricity.

As Titan travels around Saturn, the planet will show phases. Because Saturn occupies a much larger fraction of Titan's sky than our Moon does in ours, on the Saturn-facing hemisphere of Titan it will never quite get dark except during an eclipse. Roughly, Saturnshine is about one-thousandth as bright as the Sun on Titan. In addition to causing scattering in the line of sight, the fact that Titan's haze also bounces light beyond the terminator means it will create a long, bright twilight.

Stratospheric circulation

Prior to the advent of near-infrared telescopes and Cassini, most of our observations of Titan penetrated only to its stratosphere, where the long column of haze defines its appearance in visible light and molecular absorptions give prominent signatures in the thermal infrared.

The north–south asymmetry in the haze is a seasonal effect caused by the meridional winds in the upper atmosphere which transport the haze in a large convection cell from one pole to the other. For much of each half-year, the motions would be rising in the summer hemisphere and descending in the winter hemisphere. This would drag some of the haze across the equator to create the detached haze layer as the upper branch of the cell, with the accumulation causing the winter hemisphere to progressively darken. This picture, indicated with numerical models prior to Cassini's arrival, also explained the connection between the polar hood and the layers of detached haze seen by Voyager.

In the downwelling (winter) branch of the circulation, air which is laden with photochemical products from above is heated by being compressed in its descent, and paradoxically at high altitude the air warms over the winter pole. But the lack of sunshine in the polar winter has a larger effect on the heat balance (especially lower in the stratosphere) and so the atmosphere cools there. The combination of cooling and abundant condensable material makes peculiar stratospheric clouds, not dissimilar to the polar stratospheric clouds that form high in Earth's atmosphere, also known as 'noctilucent' clouds. The presence of the organics in both gas and aerosol form allows these regions to radiate heat away efficiently, causing even more cooling.

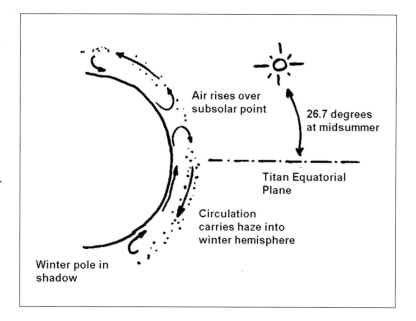

ABOVE A schematic of the meridional (Hadley) circulation and its effect on the haze. *(Author)*

BELOW This T57 profile of the atmosphere of Titan at the end of polar winter shows a dramatic deviation from the 'normal' equatorial profile. The tropopause is a couple of degrees cooler and there is some warming over the 50–70km range (adiabatic heating of downwelling air). Most remarkable is the strong cooling through much of the stratosphere in the 7–200km range. *(Author)*

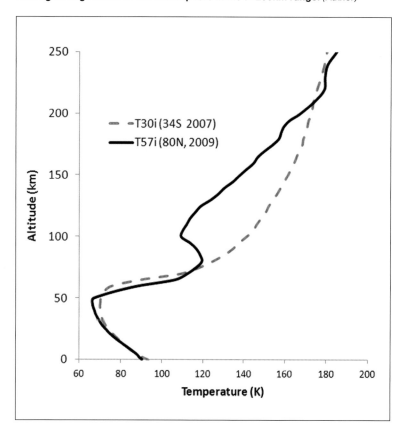

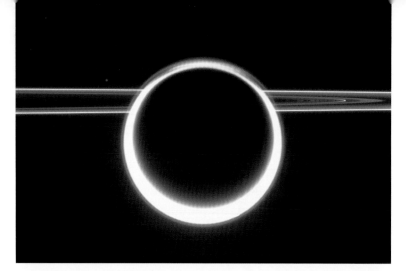

LEFT **A backlit view of Titan in 2006, with the rings and little moon Janus in the background. The thin detached haze layer can be seen all around Titan, and multiple layers of haze swirl in the northern polar region.** *(NASA/JPL/Space Science Institute)*

ABOVE **A false-colour VIMS image in 2006 showing what appears to be an ethane cloud swirling in the stratosphere. The downwelling of ethane-rich air in the northern summer led to supersaturation and the creation of ethane droplets.** *(NASA/JPL/U. Arizona/LPGNantes)*

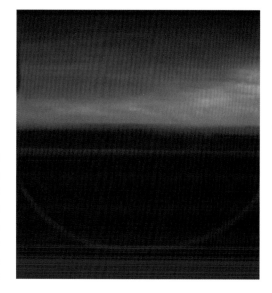

RIGHT **Noctilucent clouds of ice crystals can occasionally form high in Earth's stratosphere, somewhat analogous to Titan's various clouds.** *(John Turtle)*

The fiendishly complicated coupled system of hazes, gases, temperature and wind produces very dramatic changes in the polar regions. Although Cassini observations of the haze (by all the optical instruments) and gases and temperatures (mainly by CIRS) documented the pieces of the processes, a fully integrated picture of these seasonal changes remains to be assembled.

Upon Cassini's arrival in 2004, which was southern summer for Titan, the downwelling was firmly established in the northern hemisphere. There were multiple layers of haze around the north pole (perhaps with different compounds condensing at different altitudes, one of them evidently ethane).

A puzzle was that the detached haze layer had been measured by Voyager to be at an altitude of 350km, yet Cassini's images showed a pronounced layer at 500km. Was there a mistake, or were they different layers, or had a single layer altered altitude? The extended observations by Cassini revealed that they were indeed the same layer, and that the changing circulation pole-to-pole basically allowed the layer to collapse down from the Cassini/summer 500km level to the Voyager/spring 350km level just around equinox.

Particularly remarkable is the evolution of nitrile compounds in Titan's atmosphere, such as hydrogen cyanide. These gases were known from Voyager to show significant variation with latitude and season, and Cassini was able to track these changes in great detail. As expected, they were more abundant in the north when Cassini arrived, and on through northern spring equinox (the Voyager season, where the same pattern had been seen). Then, as the circulation reversed, hydrogen cyanide began to accumulate instead in the south. What was *not* expected though, was the appearance of a dramatic discrete cloud of hydrogen cyanide ice crystals over the south pole in a 'dipole' shape.

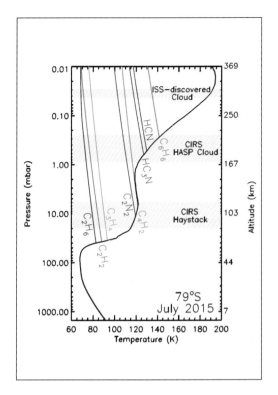

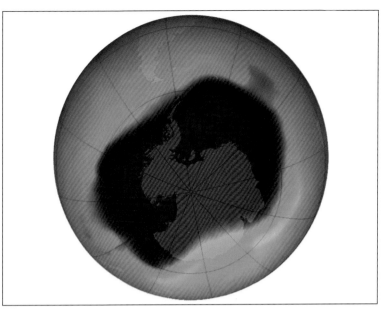

ABOVE Various compounds with different volatilities will condense at different altitudes on Titan (where the respective 'dew point' saturation curves meet the temperature profile). This gives some clues as to the composition of some of the observed clouds. *(C. Anderson/NASA)*

ABOVE The Earth's ozone hole has some commonalities with Titan's polar hood. It is dynamically isolated from the rest of the atmosphere by a circumpolar vortex and the exotic chemistry in winter darkness creates a quite distinct composition. *(NASA)*

BELOW Stratospheric zonal winds estimated from thermal gradient measurements by the CIRS instrument at intervals of 2–3 years throughout the Cassini mission. The circumpolar jet at mid-high-latitudes starts in the north (in winter), weakens after the equinox in 2009, and then forms around the south pole. *(R. Achterberg/NASA GSFC)*

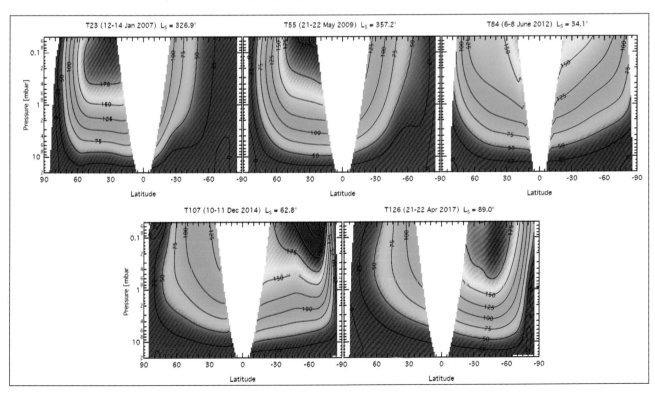

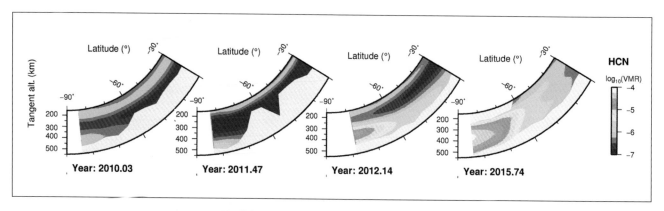

ABOVE The build-up of hydrogen cyanide in downwelling air over Titan's south pole as it heads into winter. A $\log_{10}$(VMR) value of −4 corresponds to a vapour mixing ratio of 100ppm. *(Adapted from N. Teanby)*

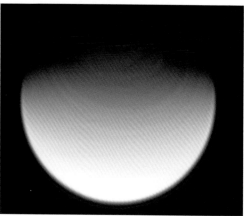

ABOVE A banded structure is evident in the north polar haze in this ultraviolet image over the north polar regions in February 2007. The faint detached haze layer can just be seen around the limb. *(NASA/JPL/SSI)*

ABOVE A banded structure is evident in the south polar haze in this colour composite image from 2013. An unsharp mask filter was used to highlight the filigree banded structure.
(NASA/JPL/SSI/Gordon Ugarović)

BELOW A near-infrared image by Cassini in December 2012. The methane band (889nm) filter detects only the upper part of the haze (slightly darker in the southern hemisphere), and dramatically highlights the peculiar polar vortex cloud that is standing proud over the south pole. *(NASA/JPL/SSI)*

BELOW Colour detail looking down on the south polar vortex cloud as those latitudes enter winter darkness in 2012. The bi-lobed structure in many ways resembles that of a polar vortex structure seen on Venus. *(SSI/NASA/JPL)*

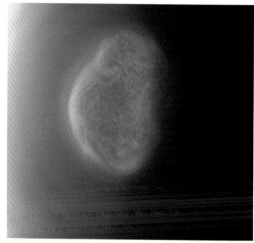

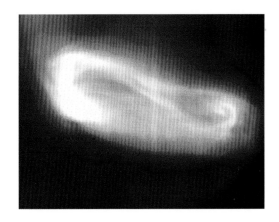

ABOVE An infrared view by Venus Express of that planet's north polar vortex. Clearly the physical and chemical conditions there differ to those on Titan, but the resulting dynamics and morphology are strikingly similar. *(ESA)*

BELOW A natural colour image of Titan in visible light by the wide-angle camera of the Cassini spacecraft on 31 March 2005, from approximately 33,083km. It looks toward the north polar region on the night side. A portion of its sunlit crescent is visible on the right. *(NASA/JPL/SSI)*

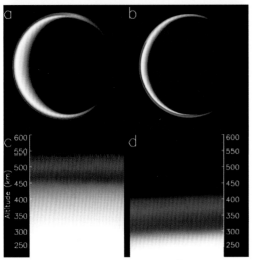

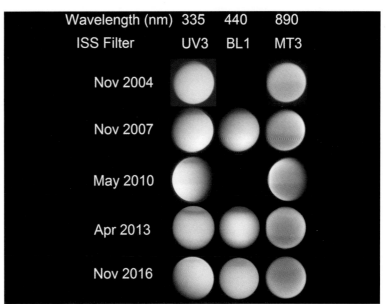

ABOVE This compilation shows the seasonal evolution of the haze observed by Cassini. The 'classic' north–south asymmetry is apparent at blue wavelengths, and reversed in the 889nm methane band (MT3). Note the multiple bands of shading, and the polar hood and collar visible in the ultraviolet and blue. *(Remcook)*

BELOW LEFT Two Cassini ultraviolet (UV3, 343nm wavelength) images – left from 2006, with the north pole at a 32° angle clockwise from vertical, and right from 2010, several months after equinox, with north near vertical. These high-phase images show the detached haze well, like those from Voyager (also just after equinox, where it was at an altitude of ~350km). Note that the altitude of the haze layer has dropped dramatically as the meridional circulation halted at equinox and began to reverse. *(Bob West/NASA/JPL/SSI)*

BELOW In this record obtained using a 21in telescope at Lowell Observatory in Arizona, the 14.5-year albedo cycle of Titan does not quite repeat. The season (SSS = Southern Summer Solstice, NSE=Northern Spring Equinox, etc.) is indicated by the vertical lines. At the 2009 NSE, Titan was 0.05 magnitudes (i.e. ~5%) darker than at the same season in 1980. *(Author, data from W. Lockwood)*

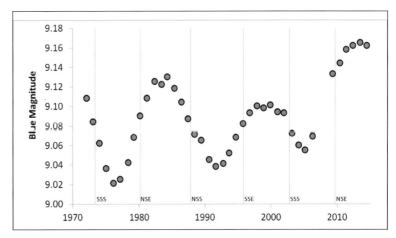

Chapter Eight

Chemistry and space interaction

No world is an island, and Titan has rich interactions with its environment in space. Ultraviolet light and charged particles drive an exotic chemistry in its atmosphere, spiced up with materials from elsewhere in the Saturnian system.

OPPOSITE Cassini looks out at a crescent Titan across the edge-on rings and the dark surface of Enceladus. Three of Enceladus' bright plumes are jetting icy grains into space, some of which reach Titan and drive oxygen chemistry in its atmosphere (the Titan/rings/Enceladus view is a real Cassini image but the plumes have been artficially brightened and Titan coloured using other image data). *(NASA / JPL / SSI / Thomas Romer / Gordan Ugarković)*

Space is not quite a vacuum, and objects in space are subject to many environmental effects that can have a lot of energy, even though they do not have much actual matter. Even a sand-grain-sized meteoroid can strike with such speed as to form a small crater, and charged particles from the Sun, or even from distant supernovae, can lance through spacecraft electronics and produce computer errors. And unshaded by an ozone layer, the harsh ultraviolet light and X-rays from the Sun can ionise molecules and break chemical bonds.

The space environment has a profound influence on Titan, and the ultraviolet flux in particular drives the chemistry which is the principal backdrop to its atmospheric evolution. But charged particles play an important role too.

Because charged particles respond not just to gravitational fields but also electric and magnetic fields, their dynamics can be very complex. And because Titan lacks a magnetic field of its own – Voyager established an upper limit of about 6 nanoteslas, some 10,000 times weaker than that at the Earth's surface – the particle flows in which it is immersed are relatively uniform across Titan's disc. Saturn, on the other hand, has an internal dynamo field which, like those of Earth or Jupiter, ducts particles along magnetic field lines to concentrate them in a circumpolar ring where they create annular auroral glows.

The plasma flow in which Titan is usually immersed is trapped by Saturn's magnetic field. This field rotates with the planet over a period of 10.5hr, and the particles being swept around with the field impinge on Titan's trailing side. Interestingly, this material includes traces of oxygen and water hosed out by the plumes of Enceladus. The plasma flow plucks some of the gas from Titan's upper atmosphere. These relatively heavy 'pickup ions' are mostly swept along with the plasma flow, streaming ahead of Titan in its orbit. In contrast, light ions (hydrogen) respond to the electric field, and are accelerated away from Saturn. These charged-particle losses account for a loss of gas from Titan of several tonnes per day.

Although the plasma environment at Titan is usually dominated by Saturn (as just described), the outer boundary of this dominance, the edge of Saturn's magnetosphere, is usually situated just outside the moon's orbit. But when the solar wind – the faint stream of charged particles emerging from the Sun – is particularly strong, its pressure pushes the boundary inward, like blowing on a bubble, and for a brief part of its orbit Titan is exposed directly to the solar wind. At least one of Cassini's encounters took place under such conditions. These and other plasma flow changes may be responsible for exciting the Schumann resonances on Titan, hinted at by measurements on the Huygens probe.

Ultraviolet light and magnetospheric particles not only sever chemical bonds, they are also able to excite atoms and molecules into more energetic states. This energy may be released again as optical or infrared emission (fluorescence or phosphorescence) at specific wavelengths. Earth's oxygen-rich and nitrogen-rich atmosphere has a prominent 'airglow' at green wavelengths. In the case of Titan, methane fluorescence and a carbon monoxide emission are strong in the near-infrared, but other emissions have been detected too.

On Earth, excitation by charged particles is enhanced at places where the particles are ducted along magnetic field lines. This gives rise to aurorae such as the Northern Lights. Saturn has aurorae too (evident in Hubble ultraviolet images as well as Cassini's data), but because Titan does not have a strong magnetic field it cannot concentrate emissions in this manner.

The energy of a charged particle interaction can sometimes impart kinetic energy to a neutral atom. If atoms are sufficiently accelerated they will barely feel Saturn's gravity, and because they do not sense magnetic and electric fields they will travel in straight lines, in that respect resembling photons of light. An instrument on Cassini was designed to image these energetic neutral atoms and discern the distribution of energy deposition.

The higher up in Titan's atmosphere, the less gas there is, and therefore the less inertia against any kind of change. So the ionosphere is subject to variation not just with season and latitude, but also with the time of day.

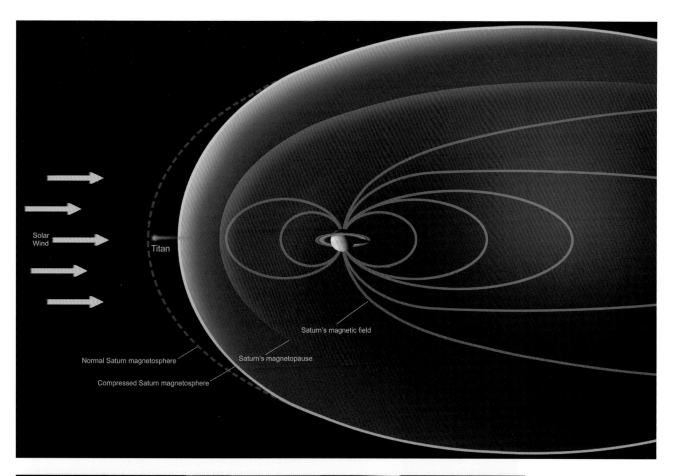

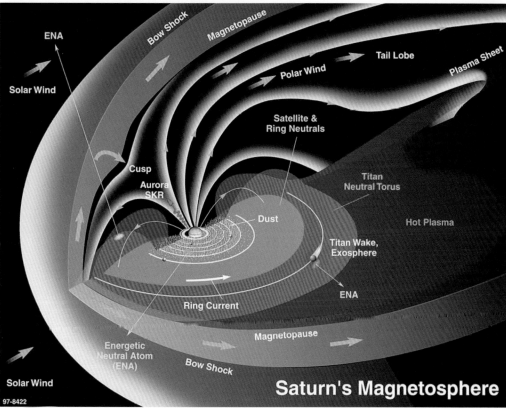

ABOVE Gusts in the solar wind push Saturn's magnetosphere inward, leaving Titan naked of magnetic protection. *(NASA)*

LEFT A schematic of Saturn's magnetosphere. A ring-current of plasma is created by the rapidly rotating magnetic field. Slowly moving Titan leaves a wake. Charged particles can exchange energy with atoms in the moon's upper atmosphere. Not being sensitive to magnetic or electric fields, these energetic neutral atoms (ENA) travel in straight lines. *(APL/NASA)*

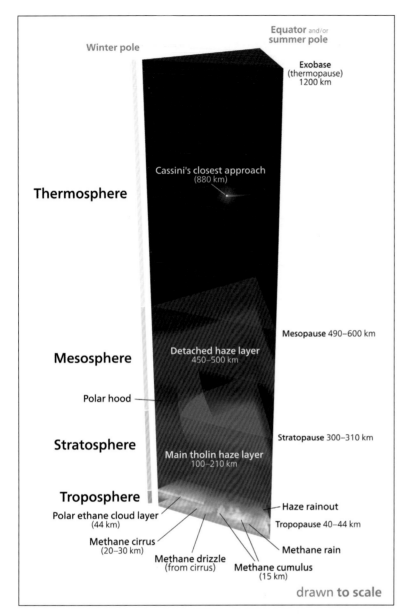

Furthermore, because the thin gas intercepts very little sunlight, it is proportionately more responsive to other energy sources, notably the particles flowing in Saturn's magnetosphere. As a result, at altitudes of 1,000km or more, the atmosphere may be influenced by the plasma conditions, such as whether it lies in the wake, or even whether the solar wind has been strong enough to push the edge of the magnetosphere inward, to briefly expose Titan to the direct flow of the solar wind (normally its orbit is completely within the Saturn magnetosphere). And then there is the other 'seasonal' effect, arising from Saturn's eccentric orbit around the Sun. For almost the entire course of Cassini's time in the Saturnian system, the planet was receding from the Sun, being 9.05 AU in 2004 and 10.06 AU in 2017. During this time, the sunlight reduced by 20%. This

BELOW On the left is a remarkable long-exposure (560secs) image of Titan while it was in eclipse on 7 May 2009, made possible only by Cassini's extremely stable pointing. The stars form short streaks due to the motion of the spacecraft relative to the moon. The lower part of Titan is faintly illuminated by light scattered around the edge of Saturn. The right-hand image was enhanced to remove the Saturnshine. This shows not only that some light still seems to be coming from Titan itself, but also that some light appears to originate from altitudes as high as 1,000km (denoted by the dashed line). *(JPL/NASA/SSI).*

ABOVE Layers of Titan's atmosphere to scale. *(KelvinSong/ wikipediaCC/BY-SA/3.0)*

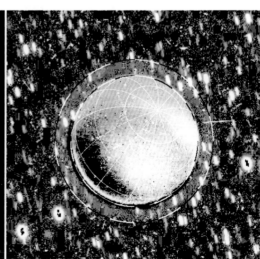

LEFT This false-colour view from Cassini's VIMS instrument at Saturn-arrival in July 2004 shows the glow from Titan's extensive atmosphere. The red channel is emission at 4.7µm from carbon monoxide which, remarkably, can glow even on the nightside. Titan seems much larger in the green channel because the fluorescence of methane at 3.3µm extends to an altitude of 700km. Light that reaches the surface is shown in the blue channel. *(NASA/JPL/U. Arizona)*

ABOVE This night view of Canada from the International Space Station illustrates the localised high-altitude red and green aurora borealis emission on the left, while the pervasive airglow (green and red together, appearing yellowish) is seen as a lower-altitude arc across the image. The Manicouagan impact structure at the lower-right reminds us of space's occasionally catastrophic intrusion into the terrestrial environment. *(NASA/JSC/ESRS)*

RIGHT Seasonal change in the ionosphere of Titan observed by the INMS on Cassini. As the moon moved farther from the Sun between T18 (2006), T48 (2008) and T83 (2012), the upper atmosphere cooled and the nitrogen abundance declined. The CH_4 profile declined much more, and to lower altitudes, because the ultraviolet in sunlight that destroys methane was increased during the solar maximum. *(J. Westlake)*

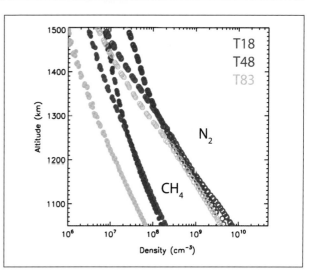

allowed Titan's upper atmosphere to cool and contract. Furthermore, the 11-year solar cycle caused the ultraviolet flux to decline in the late 2000s and then climb again in the 2010s. By destroying methane in the upper atmosphere more effectively, the higher ultraviolet flux caused the methane to decline more than the background nitrogen gas.

Although ultraviolet light and most magnetospheric particles are absorbed high up in Titan's atmosphere (~1,000km), some energy does make it to lower altitudes, as evidenced in profiles of the ionosphere. Small meteors might also cause sporadic deposition of material, and ionisation, at altitudes of several hundred kilometres.

The most penetrating charged particles are those with high mass and energy. Cosmic rays are atomic nuclei accelerated to a substantial portion of the speed of light by various astrophysical processes far beyond our solar system, including supernovae. On interacting with matter they cause mayhem at the atomic or subatomic scale. In particular they can damage microelectronic circuits (corrupting the computer instructions by flipping digital bits in spacecraft memories), and produce reactive free radicals in the water-rich tissue of living things (corrupting genetic code). More generally, however, these particles just strike an air molecule, or a rock. Often, the encounter may just tear off a few electrons from its victim, causing ionisation. As a cosmic ray coming in from space gets deeper into the atmosphere, it becomes progressively more likely to strike a molecule, and so the ionisation due to cosmic rays increases from zero at the top of the atmosphere. Eventually fewer and fewer of the particles will survive, and so the ionisation rate will fall again once a given density of matter has been traversed. On Earth, this corresponds to an altitude of ~20km. But Titan's thicker atmosphere means the maximum ionisation occurs at ~60km (predicted prior to Cassini, this was confirmed by Huygens measurements of electrical conductivity during its descent through the atmosphere).

If a cosmic ray makes a lucky strike on an atomic nucleus, this will create a spray of neutrons and other particles and transmute the target atom (or others caught in the neutron

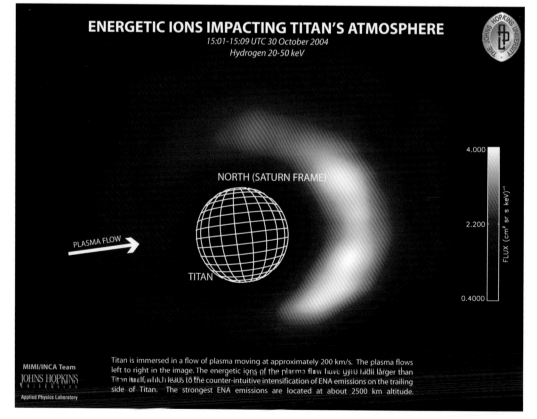

RIGHT Because energetic neutral atoms (ENA) are not sensitive to magnetic or electric fields, they travel in straight lines. Images from Cassini's MIMI instrument showed where the particles originated. Counterintuitively, because the plasma that excites neutral atoms spirals rapidly around the field lines of Saturn's magnetosphere, the ENAs are produced primarily on the side of Titan opposite that exposed to the bulk plasma flow. *(APL MIMI Team)*

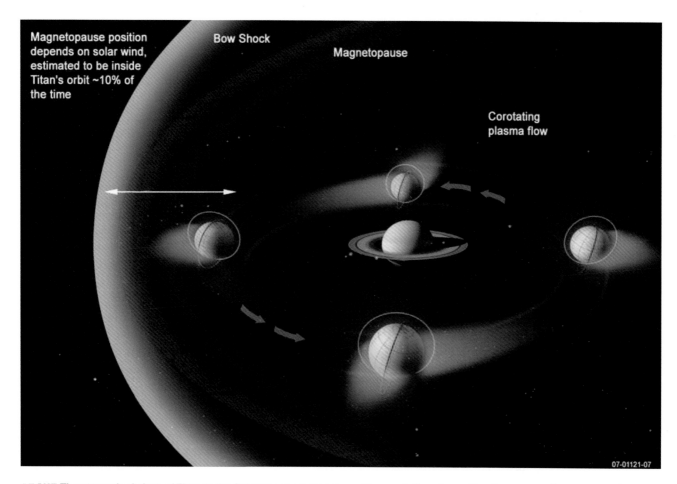

ABOVE The atmospheric loss of Titan to the Saturnian magnetosphere. The corotating plasma flow (grey arrows) swept around by the planet's magnetic field can 'pick up' heavier molecules like methane (yellow tails), whereas hydrogen ions are swept outward (green tails) by a radial electric field. *(APL)*

BELOW The ionosphere of Titan, measured by the Cassini radio occultation experiment on the T31 flyby. The thick layer 1,000–1,500km is persistent, but the lower (sporadic?) layer around 600km is absent in the other 20 or so profiles. *(Author)*

BELOW Electrodes on the Huygens probe measured the electrical conductivity of the air to determine the profile of electrons in Titan's lower atmosphere. The peak at 60km is the result of cosmic ray ionisation. *(Author)*

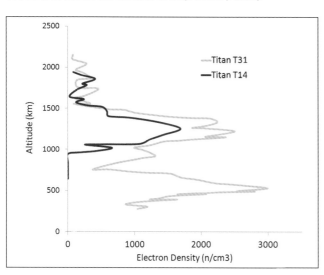

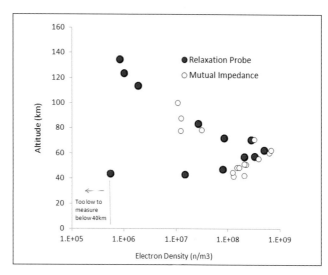

shrapnel) into other elements or isotopes. The radiation from this process (in the form of neutrons and gamma rays) can be used to diagnose the composition of asteroids or the Moon and Mars but Titan's thick atmosphere prevents the cosmic rays from reaching the ground. One effect they might have though is the conversion of N^{14} atoms into C^{14}, known as radiocarbon. On Earth, this takes the chemical form of carbon dioxide, and is respired (like the much more abundant 'regular' carbon dioxide) by living things. When things stop respiring, the proportion of their carbon that is radiocarbon is like that of the air. But the radiocarbon decays with a half-life of about 7,000 years. This gives us a way to calculate how long ago something died, because the proportion of radiocarbon in a measured sample will decline from a known initial value at a well-determined rate. On Titan, radiocarbon is probably incorporated into the haze, possibly making it very slightly radioactive – perhaps a future mission will be able to determine how 'old' the surface organic deposits are. Apart from this trace of radiocarbon (and maybe a little potassium in salts dissolved in surface ices) the surface environment should have a very low background radiation. Nevertheless, some of the most energetic cosmic rays will get through (the atmosphere has a column mass about the same as 100m of water on Earth) and possibly drive some slow chemical change on the surface.

Photochemistry

The biggest contribution of energy to the upper atmosphere is solar ultraviolet radiation. Its main effect is to break the carbon-hydrogen bonds in methane and other hydrocarbons. The resulting fragments recombine to form heavier hydrocarbons and molecular hydrogen (H_2).

Molecular nitrogen (N_2) is more abundant than methane but its triple bond is much harder to break, being susceptible only to the (much less abundant) hard ultraviolet light, or energetic particles in Saturn's magnetosphere. However, the fact that at least some nitrogen atoms are introduced into the chemical mix makes Titan's photochemistry much richer than that of the giant planets.

Although N_2 and CH_4 are heavy enough that their molecular speeds at Titan temperatures are low compared with the escape velocity (such that its gravity can hold onto these gases for the age of the solar system), molecular hydrogen is able to escape. Its presence in the atmosphere today is the result of replenishment by photochemistry.

A similar argument applies to methane. Since it is being continuously destroyed by sunlight, the amount in the atmosphere today would be destroyed at present rates in about 10 million years. Its continued presence suggests that some process or reservoir must be resupplying it. Prior to Cassini, it had been imagined that surface seas of methane might be such a reservoir, gently evaporating to compensate for the photochemical destruction (itself limited in part by the number of ultraviolet photons in sunlight). Multiplying this photon flux by the age of the solar system indicated that perhaps several-hundred-metres-worth of liquid methane could have been used up.

In the early days of Titan science the atmosphere was viewed rather one-dimensionally, with altitude as the only variable that really mattered. Methane would diffuse up from the lower atmosphere, be broken up by ultraviolet light, and the fragments would recombine to form a variety of compounds. As these diffused downward, they might occasionally react with each other to create yet larger compounds, and when sufficient of them had built up and/or mixed down to lower levels, where it was colder, they would first condense to form the hazy clouds and eventually be deposited on the surface. The situation was similar to photochemistry that might be imagined to occur in the atmospheres of the giant planets – apart from the fact that while their gravity is enough to retain the hydrogen produced by methane breakup, on Titan this escapes to space.

Reality, of course, was rather more complicated. Nitrogen is abundant on Titan and makes the chemistry a little different from the hydrogen-rich giant planets. Although the N_2 bond is not easily broken by ultraviolet light, some of the electrons circulating in Saturn's magnetosphere have energies sufficient to add excited nitrogen atoms into the mix, making for

CARBON AND HYDROLOGICAL INVENTORIES OF EARTH AND TITAN

	Reservoir	Inventory (Cubic km)	Inventory GigaTons Carbon	Column Mass kg/m²	Pressure mbar
Titan	Methane in Atmosphere		360,000	4,000	6
	Ethane/Methane Lakes	>70,000	30,000	*400*	0.6
	Sand Dunes	~100,000	80,000	1,000	2
Earth Carbon	Gas		140		
	Oil		300		
	Coal		3,500		
	Atmosphere		720		0.4
	Biosphere		2,000		
	Ocean		38,000		
Earth Water	Atmosphere			20	~2
	Oceans	1.3 Billion		2,400,000	240,000
	Lakes	125,000			
	Ice caps	>30,000,000			

a much richer range of compounds. Several factors might change over time. Firstly, the methane might run out, ending the production of hydrogen (which would escape to space) and organics (which would settle out), leaving a clear (blue) atmosphere of just nitrogen. Secondly, the ultraviolet flux may vary, and if it falls to low values (as will happen as the Sun evolves to become a red giant) there may be a similar effect, except the increased solar output will warm the surface, perhaps to the point of melting the ice and sweating the 'old' organics into the atmosphere as greenhouse gases and/or clouds. Finally, the magnetospheric electron energies or flux might change as a function of the evolution of the solar wind, or Saturn's magnetism, or even of the rings. Such changes might affect the balance of carbon and nitrogen in Titan's haze. Recently, scientists got a view of just such an 'alternate Titan' photochemistry with the flyby of the New Horizons mission at Pluto, which was revealed to possess an impressive suite of haze layers.

Laboratory experiments designed to recreate Titan's chemistry and haze formation process are not unlike the Miller-Urey experiments which explored the early Earth and the possible role of atmospheric chemistry in the origins of life. These

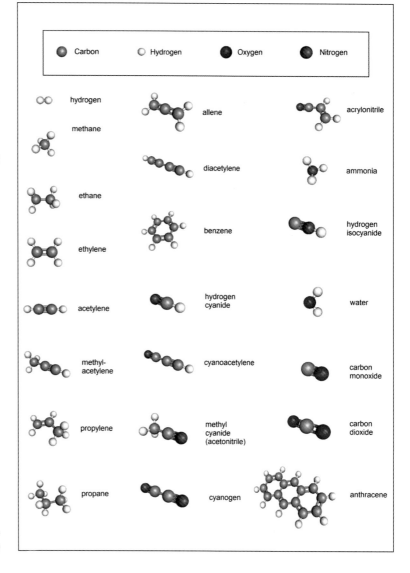

RIGHT Molecules identified in Titan's atmosphere.

(Author)

RIGHT A schematic of Titan's atmospheric chemistry leading to the formation of the solid material known as tholin. *(NASA/SwRI)*

BELOW Lab experiments matched the far-infrared ('vibrational') spectral signature observed by Cassini with high-molecular weight material containing aromatic hydrocarbons that include nitrogen, a subgroup called polycyclic aromatic nitrogen heterocycles. *(NASA/Goddard/JPL-Caltech)*

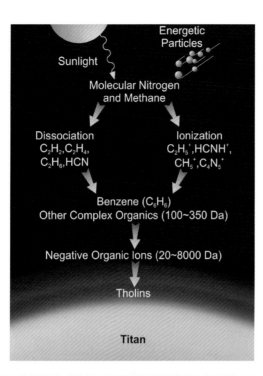

experiments, basically recirculating a mix of ammonia, methane and water and supplying energy in the form of an electrical spark, create a rich soup of organic materials, including amino acids. The Titan equivalents use nitrogen and methane, and various energy sources such as a hot plasma (like a spark), cold plasma (rather like the glow of a neon lamp), ultraviolet light, or even focused laser pulses. These all have the effect of breaking the nitrogen and methane apart (sometimes chemists use tricks like adding traces of benzene or acetylene to help things along), then the fragments recombine to create a wide range of chemicals, including a tarry or powdery mix collectively termed tholins (named after the Greek word for mud) which appear to resemble Titan's haze. But there are inevitable differences; for example, small amounts of oxygen that leak into the glassware can modify the chemistry, the temperature at which the experiment is performed makes a difference, the walls of the experiment vessel have an effect, and so on. Nevertheless, experiments of this type have been enormously illuminating.

Computer simulations have played a major role in understanding Titan's chemistry too. If we know the rates at which reactions occur (typically from laboratory experiments), then for each level in the atmosphere we can input the amounts of each chemical and calculate how fast the reactions between each combination should take place. These will also depend on the amount of light and energetic particles, and on the temperature, all of which vary with altitude. But the amount of ultraviolet light will also depend upon how much of various absorbing molecules is present above a given level, so the system has feedbacks. A somewhat simple exercise for some worlds, the models become fiendishly complicated for Titan because there are so many possible molecules and reactions. Then vertical transport and possible condensation must be taken into account. And if that were not enough, everything varies with season and latitude! This kind of work is commonly farmed out to graduate students.

Even with more sophisticated models, we find we are still left with the puzzle that confronted Titan after Voyager – why is there still methane in the atmosphere? The seas are not big enough to provide much of a buffer against depletion. The answer presumably lies in Titan's interior. The question of where all the methane (carbon) went is still puzzling – we expected there to be more ethane on the surface than we see, if the methane and the photochemistry it feeds has been around for most of Titan's history. But it is easy to imagine other answers to that part of the puzzle – some ethane has perhaps been swallowed up by clathrate ices in the crust, and perhaps more of the carbon ended up in heavier (solid) molecules, making up Titan's sand, for example.

Models developed prior to Cassini had about 100 different chemical reactions for compounds with up to 6 carbon atoms; those heavier than that were regarded as 'soot'. Cassini revealed a wide variety of compounds heavier than C_6 which extended much higher into the atmosphere than expected.

Remarkably, the INMS instrument didn't just detect a few simple molecules such as N_2, CH_4, C_2H_6, C_2H_2 and so on with molecular masses of 30 or less, it yielded a complex mass spectrum of compounds with regular peaks declining only slowly for molecular masses up to 100 (as far as the instrument could go). The CAPS instrument also found negative ions that indicated an abundance of materials with very high molecular weights (up to 10,000). These may well have been polycyclic aromatic hydrocarbons (PAH) and their nitrogen-substituted equivalents. One PAH that seems particularly abundant is anthracene ($C_{14}H_{10}$) with three carbon rings in a line and a molecular weight of 178. It occurs on Earth in coal tar. The very largest molecules with hundreds of carbon atoms

LEFT A faint violet discharge is seen in a 'cheap and dirty' room-temperature tholin production experiment in Arkansas. The box underneath is the high voltage power supply. *(Author)*

LEFT A more elaborate tholin investigation at the University of Arizona in which the experiment is submerged in a dry-ice freezing mixture to maintain low temperatures. A dense array of pipes supply fresh gas mixtures. *(Author)*

LEFT Deposition of brown tholin tar on the glassware of a tholin experiment in Paris, France. Notice that the discharge conditions and gas mixture in this setup cause a red, rather than violet, emission. *(P. Coll)*

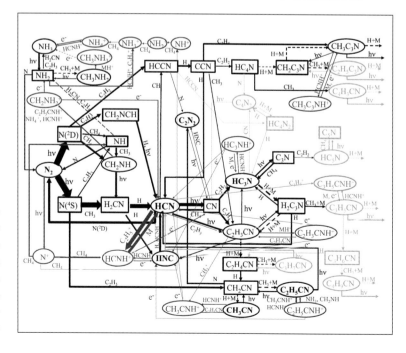

RIGHT A part of a network of reactions in a typical modern photochemical model. This only shows the nitriles, not all the hydrocarbons! Each arrow corresponds to a reaction whose rate may be dependent on temperature. And all the reactions may take place at different levels in the atmosphere. The complexity that emerges is readily apparent. *(M. Dobrojevic)*

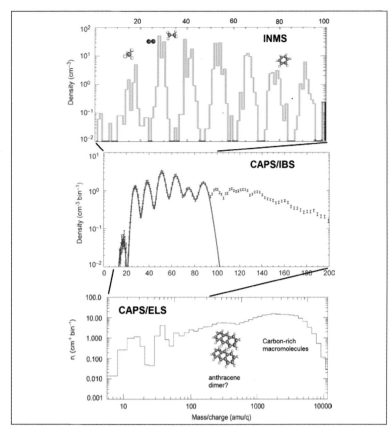

ABOVE Cassini's INMS measured abundant molecules and ions in Titan's atmosphere up to 100 mass units (i.e. up to C_7 compounds). The regular peaks in the upper panel indicate the addition of a single carbon atom. The CAPS data for positive ions (middle panel) of up to 200 units show that the abundances tail off only slowly toward higher mass. The negative ions (bottom) show that huge molecules with hundreds of carbon and nitrogen atoms are formed. *(Author)*

likely have the form of sheets and tubes. Being several nanometres across, they start to cross the boundary between macromolecule and 'particle'.

> **Organic chemistry just now is enough to drive one mad. It gives me the impression of a primeval forest full of the most remarkable things, a monstrous and boundless thicket, with no way of escape, into which one may well dread to enter.**
>
> **Friedrich Wohler** in a letter to J. J. Berzelius, 28 January 1835

When we look at tholins in the laboratory, we find that the material is present in little blobs a few tens of nanometres across, which in turn form clusters, aggregate particles about a micron across. It had been suspected (shortly after Voyager) that Titan's haze might be made of fractal aggregate particles, not only because it appears physically natural for such shapes to form, but the optical properties of the haze (if assumed to be of spherical

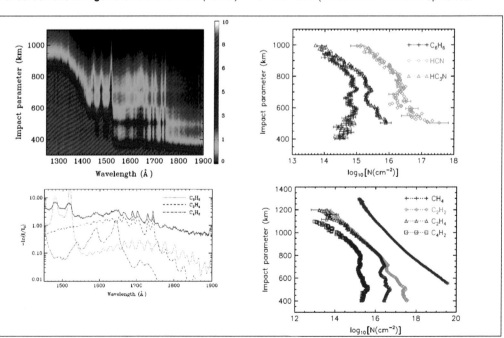

RIGHT Ultraviolet observations of stellar and solar occultations by Cassini's UVIS were able to profile haze and certain molecules in the 400–1,000km altitude range for Titan, where few other observations were possible. Analysis was able to disentangle the overlapping absorptions of the different gases in order to produce abundance profiles. *(Author, from data by T. Koskinen)*

TABLE: COMPOSITION OF TITAN'S ATMOSPHERE

Common (official) Name	Formula	Amount (in stratosphere unless otherwise indicated)
Nitrogen	N_2	95% near surface
		~98% in stratosphere
Methane	CH_4	4.9% at surface near equator
		1.4% in lower stratosphere
		~2% at ~1,000km
Hydrogen	H_2	0.1–0.2% in lower atmosphere
		~0.4% at ~1,000km
Argon	Ar^{40}	43ppm (produced by K^{40} decay)
	Ar^{36}	28ppb
Ethane	C_2H_6	~20ppm
Carbon monoxide	CO	~45ppm
Acetylene (ethyne)	C_2H_2	3.3ppm
		19ppm at ~1,000km
Propane	C_3H_8	700ppb
Hydrogen cyanide	HCN	800ppb in winter stratosphere
		~100ppb in summer stratosphere
Ethylene (ethene)	C_2H_4	160ppb
Carbon dioxide	CO_2	15ppb
Methyl acetylene (propyne)	C_3H_4	10ppb
Acetonitrile	CH_3CN	a few ppb
Cyanoacetylene	HC_3N	5ppb in winter stratosphere
		<1ppb in summer
Methyl acetylene	CH_3C_2H	5ppb
Cyanogen	C_2N_2	5ppb
Water vapour	H_2O	8ppb
Diacetylene (buta-1,3-diyne)	C_4H_2	1.5ppb (slightly higher in winter)
Benzene	C_6H_6	1.4ppb at winter pole, <0.5ppb elsewhere
Anthracene	$C_{14}H_{10}$	trace
Note: 'ppb' means parts per billion.		

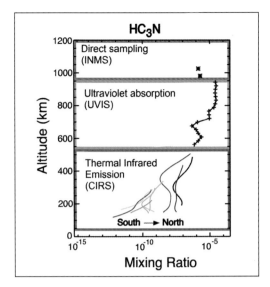

ABOVE The vertical profile of cyanoacetylene in Titan's atmosphere was measured at different heights by three different Cassini instruments. (They do not quite match up because they were made at different latitudes and seasons; the latitude variation is shown in the CIRS profiles at the bottom.) *(Author, adapted from Sarah Horst)*

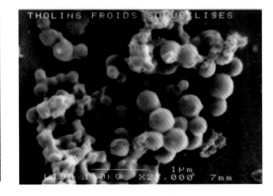

BELOW An optical image of tholin particles in a study of the effectiveness of non-stick coatings for the windows of Dragonfly's cameras. *(J. Benkoski/APL)*

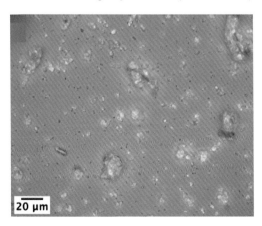

BELOW A coloured ultraviolet view from 2004 of haze layers at the north pole of Titan. *(NASA/JPL/ESA/SSI)*

ABOVE An electron micrograph of a tholin. The individual 'monomers' are spheres a fraction of a micron in diameter, but they clump together to form irregular fractal aggregate particles about a micron across. *(P. Coll)*

ABOVE Pluto's tenuous atmosphere of nitrogen, methane and carbon monoxide has its own flavour of tholin haze layers. Pluto's chemistry is almost entirely driven by ultraviolet from the distant Sun. The surface pressure of about 10 microbars is comparable to that for Titan at an altitude of 400km. *(NASA/APL/SwRI)*

particles) measured in different ways were mutually incompatible. Aggregates of small particles are more effective at blocking blue or ultraviolet light, compared to that at red wavelengths, than spherical particles with the same red cross-section.

Oxygen chemistry

In the 1980s and '90s, scientists (including myself) puzzled over why Titan's atmosphere has so much carbon dioxide – although it was just a trace, it was more than expected. The surface and lower atmosphere are so cold that CO_2 would be trapped there – its vapour pressure is just too low, just as there is not much rock or iron vapour in the air we breathe. So the CO_2 observed at high altitudes had either to be produced there or brought there from outside. The original idea was that little icy meteors would be continuously 'burning up' (or at least evaporating) as they sprinkled in from space, and the water vapour they liberated would enter the photochemistry, reacting with the methane, carbon monoxide and so on to produce CO_2. This seemingly exotic introduction of matter from space actually occurs on Earth, but with rather different materials. Iron, sodium, and magnesium are not obvious elements to find in the Earth's stratosphere and mesosphere (at 80–100km altitude, the edge of space), but because these elements have strong optical signatures (even a tiny trace of sodium, like from a salt grain, will make a flame turn a violent yellow) we can detect their faint glow using lasers and telescopes on the ground, and their surprising abundance in the mesosphere is attributed to meteors burning up.

However, when we considered how much water vapour 'should' be delivered to Titan's upper atmosphere using the best information on how many meteoroids ought to be there, it was not enough. We tried some modelling tricks like accounting for secondary ejecta (a meteoroid hits an icy moon like Dione, say, and the ice shrapnel could multiply the amount of material in this part of space – the Galileo spacecraft had measured such an ejecta process at Ganymede in the Jovian system) or imagining that some of the ice, like that of comets, was already CO_2 and not just water (which would be only incompletely converted to CO_2 on Titan). But when our tricks were insufficient, we waited...

On Cassini's arrival, the answer was glaringly obvious. The tiny moon Enceladus was hosing water vapour (and salty ice grains) through vents near its south pole. Sometimes (inaccurately) referred to as 'geysers', these plumes seem to be jetting material from the watery interior. Some of this snows back as ice particles, but Enceladus' gravity is so weak that a fair amount escapes into space. However, although the ice grains escape

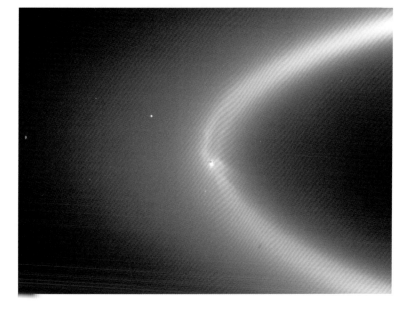

LEFT Enceladus' plumes feed Saturn's diffuse E-ring, as demonstrated in this high-phase-angle (backlit) Cassini image on 15 September 2006, at a distance of approximately 2.1 million kilometres. The forward-scattering geometry is highly favourable for detecting small particles. *(NASA/JPL/SSI)*

Enceladus itself, they remain stuck in orbit around Saturn, forming the faint E-ring, whose density is greatest at Enceladus' distance from Saturn.

And there they would remain, were it not for the rich physics of the Saturnian system. Small particles can easily acquire and retain an electrical charge – particularly in space. This makes them susceptible to electrodynamic effects. At Enceladus' distance from Saturn, an object in a circular orbit has a period of 33hr but Saturn spins in 10.5hr, as does its magnetic field. As the field sweeps past the ice grains it tugs them, increasing their energy, and making them spiral outward. Electric fields cause other effects, but the overall result is that the charged ice grains and clusters of water molecules migrate outward and some of them reach Titan. Thus many of the oxygen atoms floating around in Titan's atmosphere spent most of the history of the solar system inside Enceladus, and only relatively recently were spewed into space in a spectacular plume where they became susceptible to electromagnetic forces!

Computer models show that this oxygen flux could permit traces of heavier molecules to form on Titan but the best way to incorporate oxygen into the chemistry is to do it on the surface, as we shall see later.

Dance of gravity

Titan is connected to the rest of the Saturnian system not just by the exchange of matter and charged particles, but also by gravity. Indeed, this is what keeps Titan in the Saturn system! Titan's eccentric orbit causes tides in its seas, atmosphere and interior, but the reason for the persistence of this eccentricity has not (yet) been determined. Perhaps it is recent, forced by some *Deus Ex Machina* event like an impact or close encounter, or maybe some exotic orbital resonance with Jupiter is responsible.

It appears to be unrelated to the fact that Titan is in a 4:3 orbital resonance with Hyperion, the next moon out from Saturn. Such resonances can have profound effects (for example, the Nice Model of the early solar system posits that Jupiter and Saturn passed through a 2:1 resonance that had the effect of putting Uranus and Neptune in their current orbits and diverted a hail of asteroids into the inner solar system), but Hyperion is far too small to produce an appreciable effect on Titan.

Yet one feature where Titan's influence is visibly demonstrated is in Saturn's rings. The Titan Ringlet inside the Colombo Gap is slightly oval-shaped and always points its long axis toward Titan. The behaviour and orientation of this ringlet are controlled by a resonance between the 16-day orbit of Titan and the rate at which the oval-shaped orbits of the ring particles 'precess' around the planet. Since this precession is sensitive to the internal gravity structure of Saturn, the Titan Ringlet is yielding useful information about the outer portions of Saturn's interior.

So Titan is just part of a richly interacting system which Cassini was well-equipped to explore during what proved to be a 13-year mission which inspected the moon Phoebe on its way into the system and encompassed the far reaches of the magnetotail (drawn out by the solar wind downstream of the Sun) to just inside the rings, and ultimately to a fiery end by plunging into Saturn itself. While some mysteries, such as the trace oxygen chemistry, seem largely solved, others remain. After Huygens' descent, and Cassini's 126 rich but fleeting flybys of Titan, the next steps in Titan exploration will focus exclusively on the moon itself.

LEFT The dark Colombo Gap in Saturn's rings contains a bright band, the 16–32km-wide Titan Ringlet, whose slightly elliptical shape follows Titan around in its orbit. At the time of this 2009 image the Sun was only just above the ring plane and causing shadows to highlight any corrugations. Also evident is a very narrow feature that transitions from bright at the top of the image to dark at the bottom. This is a bending, or vertical, wave generated by synchronicity with Titan's out-of-plane motion. *(NASA/JPL/SSI)*

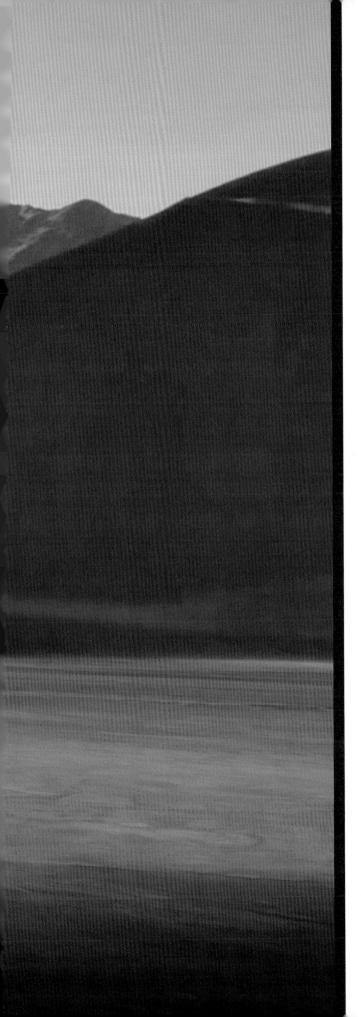

Chapter Nine

Future exploration

―⦅●⦆――――――

Titan's rich chemistry and diverse landscape demands further investigation, and the fact that it offers a means for a vehicle to move around has prompted some of the most innovative ideas in planetary exploration.

OPPOSITE Designed by APL for NASA's New Frontiers program, Dragonfly will be the next mission to Titan, arriving in 2034. This octocopter-lander uses a radioisotope generator to sustain a mission of exploration lasting at least 2.5 years that will begin in one of the dunefields and move in multi-kilometre hops to an impact crater. *(APL)*

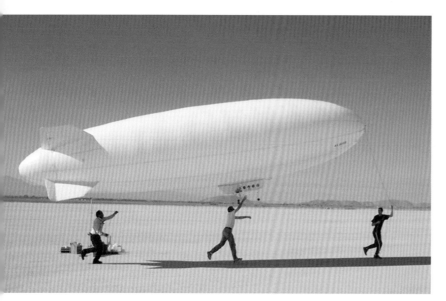

ABOVE JPL researchers flying a radio-controlled airship on a dry lakebed in California to grapple with the guidance and control challenges such a vehicle might have to confront. *(J. Hall/JPL)*

RIGHT A helicopter inspecting the Huygens landing site, advocated by the author circa 2001. The co-axial rotors allow (relatively) efficient packaging inside an entry shell and obviate the need for a tail rotor. The concept was a couple of decades ahead of its time. *(James Garry)*

Ideas for the exploration of Saturn and Titan beyond Voyager have swirled around since the mid-1970s, and even before the thickness of Titan's atmosphere was measured its importance was recognised. For example, one idea envisaged using 'aerocapture', where a spacecraft in a shaped heat shield would use the atmosphere to slow down from its interplanetary trajectory and then climb out into orbit around Titan or Saturn. French and US scientists also advocated exploration using balloons. These ideas fell into abeyance for 15 years, however, once Cassini-Huygens' development began.

In the late 1990s, with Cassini safely launched, the scientific priorities for future exploration at Saturn could only be guessed at, but NASA committees judged that Titan's surface chemistry, which Cassini-Huygens was not well-equipped to measure, would likely be top of the list.

A workshop in Houston, Texas, in February 2001 identified a number of concepts for Titan, including a rather impractical 'aerover' that featured three large spherical tyres that could be inflated with helium in order to permit balloon traverses between surface roving. While Mars exploration in the early 21st century had been dominated by rovers, uncertainty regarding the trafficability of Titan's surface has discouraged wheeled vehicle exploration there, and instead attention has focused on aerial mobility – balloons or airships – which could exploit the thick atmosphere.

A major challenge with such platforms is how to access surface material for analysis without requiring the vehicle to land. Ideas that have been kicked around (and even prototyped for lab tests) include a tethered coring penetrator such as a hollow harpoon which a hovering airship would drop and then winch back up, or a touch-and-go sampler with abrasive contrarotating wire brushes that could be dangled down a tether.

Around this time, the potential for heavier-than-air flight at Titan was recognised, because not only does the thick atmosphere help (as it does for balloons) but the gravity also makes aircraft flight easier. Aeroplanes, helicopters and even tilt-rotor aircraft were proposed. In particular, it was noted that a given helicopter configuration could hover on Titan with 38 times less power than is needed on Earth. While such a vehicle could not fly continuously given the low power-to-weight ratio of radioisotope thermoelectric generators (RTG), it could trickle-charge a large battery over a local night (lasting eight Earth days) and then fly for several hours during the day.

After Cassini's arrival, and the exciting images from Huygens, the 'Montgolfière' hot air balloon became a preferred concept – named after the Montgolfièr brothers who made the first flight in such a balloon. This platform would be less complicated than an airship, and not susceptible to leaks of helium through its large envelope (which would be cold and stiff at Titan, and would have to be packed tightly in an aeroshell for years on its way to Titan). Instead, the waste heat from an RTG could provide buoyancy. Although it lacked the horizontal control authority of a buoyant gas airship, a Montgolfière would be able to manoeuvre up and down by modulating the heat fed to the envelope or by venting hot air through a crown valve. As with recreational hot air ballooning on Earth, some horizontal control might be effected by changing altitude to levels where the winds are closer to the desired direction.

As Cassini's results emerged in 2005 and 2006, interest in Enceladus and the breadth of Titan science prompted NASA to commission 'Flagship' (i.e. Cassini-class) mission studies at these targets, and of course these funded studies attained a far higher level of technical detail than had previously been possible.

The Titan Explorer study, led by APL, emphasised a wide range of scientific investigations. A variety of architectures were considered, but the focus quickly converged on a triple-platform mission (orbiter, lander and balloon) that would be packaged for a single launch vehicle. An important constraint was that the possible radioisotope power sources were two 100W-class units: the MultiMission Radioisotope Thermoelectric Generator (MMRTG) and the Advanced Stirling Radioisotope Generator (ASRG). The latter, under development by NASA at the time, was attractive because it would use a small reciprocating engine to convert the heat from the plutonium into electrical power ~four times more efficiently than was possible for an RTG using solid-state semiconductor converters (similar to thermocouples).

BELOW An artist's concept of a Montgolfière (hot air) balloon at Titan. The tube at the neck of the balloon feeds the waste heat from an RTG into the envelope. *(T. Balint/NASA)*

BELOW This 'pocketwatch' lander envisaged by the author and James Garry circa 2007 drew inspiration from Beagle 2 and the Mars Phoenix lander. Dunes and ripples are visible in the background. A stereo camera is on a spiral mast, and a separate stick-like mast at the left carries weather sensors. A camera on a robot arm is inspecting, with light from illuminators, the ground prior to gaining a sample for analysis. A seismometer has been deployed onto the ground at the bottom-left. Because its shape is streamlined to minimise wind loads, it bears a resemblance to a computer mouse. The deflated airbags that attenuated the landing loads are off to either side. A small flapping-wing aerial vehicle is fancifully illustrated taking off at the left. *(James Garry)*

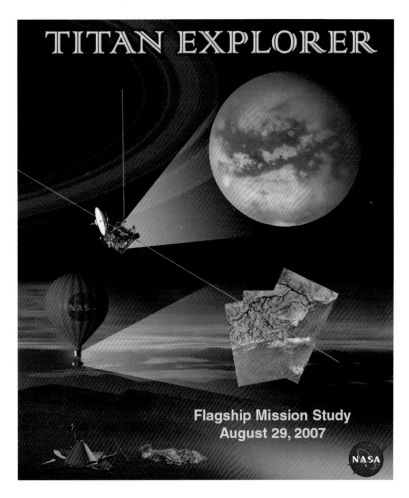

ABOVE The cover of the 197-page public version of the 2007 Titan Explorer Flagship Study. The various scales of process on Titan's surface would be studied from orbit, from a Montgolfière and from a lander. Powerful synergies arose from the combined simultaneous measurements by these different platforms. (NASA/APL)

The orbiter would use aerocapture to efficiently enter orbit around Titan. Although this would present a packaging challenge, it would give the orbiter a formidable capability: some 1,800kg, including about 170kg of instrument payload and consumables for four years of Titan operations, including propellant to conduct dipping orbits for 'aerosampling'. From orbit it would be able to completely map the surface at resolutions rather better than Cassini could during its flybys, and use a radar altimeter to directly measure the tide in the seas (which were only discovered toward the end of the study).

In order to bridge the global and local scales sampled by the orbiter and lander, a Montgolfière balloon was included to perform a regional survey. A one-year lifetime was assumed, with a float altitude of about 10km, where zonal winds should enable the balloon to circumnavigate Titan a couple of times. The 2kW of waste heat from an MMRTG is quite small, therefore a double-wall balloon envelope would be used to limit the heat loss to the cold ambient atmosphere.

When early Cassini evidence showed a remarkable diversity in Titan's surface, the question of landing site had to be confronted, at least in a preliminary ('existence proof') way. The landing site was chosen conservatively as Belet, a large equatorial region covered in organic-rich sand dunes that provides a large, scientifically appealing target with feasible landing characteristics. An airbag system would limit impact loads.

After arrival and Mars Pathfinder-like petal deployment, the lander would use a 0.5m X-band high-gain antenna mounted on a gimbal to communicate with the orbiter; this antenna would also permit a modest direct-to-Earth communications capability of 200–300 bps should that be necessary. Orbiter overflights for communication would occur in three-day groups every eight days, with each overflight separated by the 5hr orbital period.

The lander would be powered by radioisotope sources and have a near-indefinite operating capacity. Valuable synergetic measurements of the magnetic field by the orbiter and lander simultaneously would allow isolation of the induced signature of an internal water ocean on Titan; the magnetometer and a seismometer would be placed on the surface by the sampling arm to minimise lander noise. The arm would carry a microscope camera and illumination to perform spectral measurements – providing ground truth for the orbiter spectral map, which must observe through the thick atmosphere. The panoramic camera and point spectrometer would survey the landing site and support sampling operations. The site's context would be gained from a descent imager.

The option of augmenting the lander science with a small lander-launched airplane was also advocated in the 2007 study – by that time 'drones' or unmanned aerial vehicles (UAV) were finding roles in warfare on Earth. A 1kg battery-powered UAV was simply to fly from the side of the lander for a couple of hours. This concept was named 'Titan Bumblebee' because of two analogies with that insect. Firstly, the thrust to weight requirement for vertical take-off would enable the vehicle to hover, just like a bee. And then a significant part of the energy budget of a small UAV in the cold Titan environment would have to be expended to keep the vehicle warm, drawing a parallel with the specific heat management adaptations acquired by the bumblebee for the subarctic environment to which it is evolved.

Another NASA study started in 2008 for Titan Saturn System Mission (TSSM). Led by the Jet Propulsion Laboratory, this was directed to consider a Titan mission which would also study Enceladus, with optional accommodation of in-situ elements to be examined in a coordinated parallel study by ESA. A further constraint was that aerocapture was not to be considered for orbit insertion. This required that the spacecraft use electric propulsion (ion thrusters) to reach Saturn, then use chemical rockets to enter Saturn orbit like Cassini and after a resonant tour of Enceladus flybys, enter into a high orbit around Titan. This would then be gradually lowered using 'aerobraking' – a slower, gentler process than aerocapture that is now routine for Mars missions. The slower braking from a high orbit, rather than an interplanetary trajectory, saves less fuel but imposes only modest drag and heating, thereby eliminating the need for an entry shell. A 'lake lander' and a Montgolfière would be deployed during the resonant tour phase. With a launch in September 2020, planetary alignments meant that TSSM wouldn't benefit (as Cassini did and the Flagship study would) from a Jupiter flyby. As a result, the journey out to Saturn would take nine years.

Released by the orbiter on its second Titan flyby on a trajectory that would take it to Kraken Mare, a northern polar sea at about 72°N, the battery-powered lake lander was to undertake Huygens-like measurements during its parachute descent, sampling the polar atmosphere to determine any interesting differences from the low-latitude profile acquired by Huygens. The lander would splash into the hydrocarbon sea, rather as it had originally been envisaged that Huygens might. Measurements on the sea surface would include the chemical composition of the liquid, dynamics of waves, and depth of the sea. As

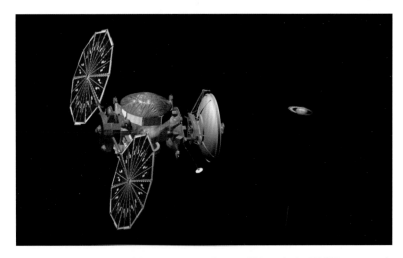

ABOVE A depiction of TSSM heading for Saturn. Although the TSSM spacecraft itself would have used radioisotope power sources, this concept envisaged large solar panels for an ion-drive solar electric propulsion stage to reach Saturn. Some more recent mission concepts, such as the OCEANUS Titan orbiter proposed in 2017, investigated using similar solar arrays without radioisotope power sources. *(NASA/JPL)*

BELOW An artist's impression of the TSSM orbiter with long ground-penetrating radar antennas and high-gain antenna deployed at Titan. The attach rings that would have held the ESA Montgolfière and lake lander (released prior to orbit insertion) are visible. At the bottom are the ASRG power units with their cooling fins. *(NASA/JPL)*

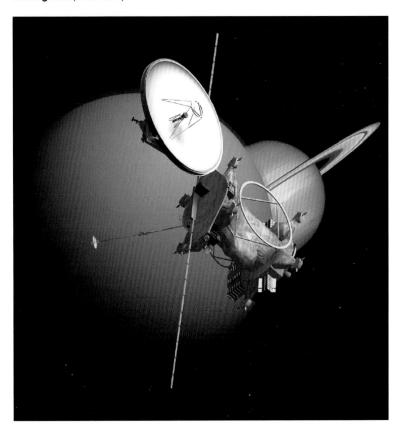

RIGHT The AVIATR proposal imagined using Stirling generators for non-stop flight to survey wide areas of Titan at high resolution for a year. *(M. Malaska)*

the mission would occur in northern winter with the only ambient illumination being diffuse twilight, imaging would be achieved using a surface science lamp. The lander would last only a few hours.

After its Saturn-orbit phase, the orbiter would use its engine to enter an elliptical orbit around Titan. A low periapsis would facilitate aerobraking and direct sampling of the high-molecular-weight photochemical products being formed in the upper atmosphere. After circularisation, the orbiter would conduct mapping, much like that envisaged in the 2007 study, although its payload would be lighter and its orbital science mission would last only two years.

However, prospects for a Flagship-class mission to Saturn receded and Europa and Mars were determined by a Decadal Survey committee to be the priorities for that class.

NASA's ongoing development of the efficient ASRG prompted an aircraft concept, AVIATR. Powered by two ASRGs, it was to deploy into Titan's atmosphere and fly continuously for a year, without touching the ground. A two-bladed propeller driven by a rare-earth-element magnet brushless direct current motor would provide propulsion. Science cameras mounted directly to the underbelly of the fuselage structure would be kept warm behind double-pane transparent windows. The flight control servo actuators would be mounted internally to the fuselage and transmit control forces and torques to aerodynamic surfaces utilising thermally insulating pushrods and torque tubes to minimise heat leaks. By flying at several metres per second (a good fraction of Titan's equatorial rotation speed) AVIATR could 'keep up with the Sun' and loiter on the dayside to maximise its communication windows with Earth, sending prolific imaging data using a dish in its nose. The concept was developed to a level of detail sufficient for a preliminary costing. At $715 million it wouldn't fit as one of the Discovery missions that NASA was soliciting in 2010 but would be a candidate for a New Frontiers-class mission.

One Discovery mission that *was* submitted was JET (Journey to Enceladus and Titan). This JPL concept would also use ASRGs, on a more or less stripped-down TSSM mission making flybys of its two targets but not entering Titan orbit. The rather austere payload comprised an infrared camera to map Titan more efficiently and at higher resolution than Cassini's VIMS, and a mass spectrometer to analyse Enceladus' plumes and Titan's atmosphere to higher molecular weights than Cassini's INMS. Unfortunately, the mission was not selected.

The only Saturn system mission to be selected by the Discovery program for further study (Phase-A) in 2011 was TiME, the Titan Mare Explorer. The technical details were developed by Lockheed Martin and APL. A large standalone capsule containing the fuel tanks and almost all of the cruise systems (thereby eliminating the requirement for a separate cruise stage) was to be launched to Titan and splash down into Ligeia Mare, the second largest of Titan's seas (and the best-mapped by Cassini at the time). It would be powered by a pair of ASRGs and would float nominally for six Titan days (96 Earth days), and possibly much longer. It would drift slowly by wind drag on its hull, camera mast and medium-gain antenna. Set to arrive in July 2023, TiME would exploit the fact that Earth barely sets below the horizon in the late northern summer on Titan, enabling data to be sent directly to Earth rather than via an orbiter.

A mass spectrometer would analyse the composition of the sea, and a sonar would profile the depth. A meteorology package would measure weather patterns and study air–sea interactions such as evaporation and wave generation. A camera would observe clouds, and as the capsule drifted toward land – perhaps after a few Titan days – the shoreline terrain too. Even with this modest payload, the mission caught the public imagination and offered a credible mission with strong scientific appeal for the available budget.

The Phase-A study fleshed out the mission details, ranging from such obvious technical issues as the thermal design for Titan's frigid (but otherwise benign) environment, procurement and test schedules, and follow-on data archiving plans. The cryogenic ocean sampling system was prototyped and tested, wire insulation tested for compatibility with liquid hydrocarbons, and splashdown tests were made using scale models to validate computer models of the loads and resurge dynamics. A major effort was the development of computer models for winds and the statistics of waves, but the fact that APL does extensive work for the US Navy in addition to NASA meant experts were on hand to calculate sonar performance and create simulations of the capsule motion in response to the wave field. But the mission was not selected for flight development.

A fanciful concept sponsored by NASA's Innovative Advanced Concepts program emerged a year or two later in the form of the Titan Submarine. Using an assumed Stirling power source, the submarine would cruise around Kraken Mare, Titan's largest

sea, around 2040-ish when it was once again summer in Titan's northern hemisphere. It would use a large electronically steered dorsal phased-array antenna to send data directly to Earth. A variation of this concept, imagined to be supported, TSSM-like, by an orbiter spacecraft overhead acting as a data relay, could exploit the radio-transparency of Titan's seas to send its data while submerged, a luxury not readily permitted for terrestrial submarines owing to the electrically conductive nature of seawater.

An interesting challenge for such vehicles is that of buoyancy control, for a couple of reasons. Firstly, the range of buoyancy required is rather large. It is only a few per cent for a terrestrial submarine, as the density

ABOVE An artist's impression of the Titan Mare Explorer (TiME). Atop the mast on the left is a planar mechanically steered medium-gain antenna to transmit data to Earth. The right mast carries a camera (in a spherical insulated housing) with an ultrasonic anemometer. *(APL)*

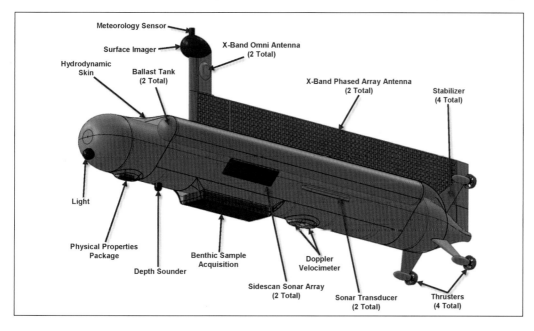

LEFT The Titan Submarine concept studied by NASA Glenn under the NIAC program revealed a number of interesting challenges in buoyancy control in the Titan environment. The slender vehicle minimiocd drag. A streamlined camera mast and a large dorsal antenna would stand above the 'waterline' when surfaced. *(NASA)*

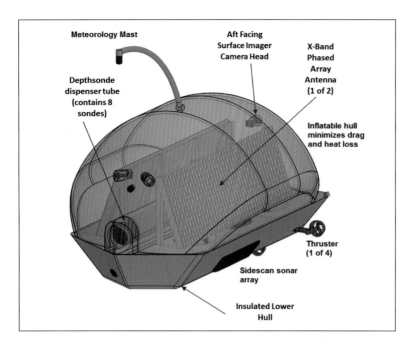

ABOVE A semi-transparent view of the upper part of the 'Titan Boat' showing the large dorsal phased-array antennas for direct-to-Earth communication. The tube for dispensing the dropsondes is in the centre, facing forward. On the hull are thrusters (brown) and a sidescan sonar array (purple). *(NASA)*

difference between warm and/or fresh water and colder salty water is quite small. In the case of Titan, however, the range of density encountered could be several tens of per cent. Thus the buoyancy tanks would have to be quite large.

A second issue is that Titan's nitrogen dissolves in cold liquid methane much more than does air in water, so it is not practical to 'blow the tanks' in quite the same way (a submarine could 'get the bends') and a piston or membrane would be required to isolate the liquid in the tanks from the displacing air.

An alternative that would avoid this problem envisaged a surface platform (i.e. a 'boat') with propulsion that would allow it to fight currents and thus stay in the safe open sea away from the shore. It would devolve the deep-sea science to expendable dropsondes that would radio temperature, composition and turbidity profiles while descending over a period of an hour or so, and possibly even take pictures from the seabed. On Earth, similar packages are known as expendable bathythermographs (XBT). Although such a concept might be more palatable to typically risk-averse space mission planners, it remains very much a 'blue sky' concept for the far future.

Remarkably, in 2016 a more immediate opportunity to consider new ideas for exploring Titan was included in NASA's New Frontiers program. These are missions in a cost class of roughly $1 billion, previous examples being New Horizons at Pluto, Juno at Jupiter, and the OSIRIS-Rex asteroid mission.

Of a dozen proposals submitted to different destinations (including the JPL-led OCEANUS solar-powered Titan orbiter), only two were selected in 2017 for Phase-A studies: CAESAR, a comet sample-return mission proposed by Cornell in partnership with NASA Goddard, and Dragonfly, a concept for a relocatable Titan lander offered by APL.

Nothing like Dragonfly had ever been seriously proposed as a NASA mission. But the 'drone revolution' of the recent decades, improved motors and motor electronics, autonomous flight using image-based terrain-relative navigation, the MMRTG, and the framing of post-Cassini science questions synergised nicely. The revolutionary concept was a lander that could spend weeks doing science at a site, then pick itself up and fly tens of kilometres in an hour or so to investigate somewhere afresh, again and again. A single vehicle would be able to carry out a science mission that would have needed a balloon and a lander in the Flagship studies of the previous decade.

Unlike the 'Titan Bumblebee', or the small Helicopter Scout intended for NASA's Mars 2020 rover, Dragonfly will not be a sub-vehicle flying from a lander 'mothership'. Rather, the local environment, with its low gravity and dense atmosphere, will enable the whole lander to lift off! At the time of writing this book (summer 2019) the Curiosity rover had driven a total of 21km on Mars during the seven years since its arrival. Dragonfly would be able to cover a greater distance in a single flight lasting less than an hour.

The vehicle will use an 'X-8' octocopter arrangement, with a square set ('quad') of over-under pairs of rotors, for several reasons. Firstly, the vehicle has to fit inside a circularly symmetric aeroshell and heat shield for hypersonic entry from space, and a quad configuration provides the biggest unblocked rotor area that fits in a circle, thereby making it most efficient. Also, the dual-quad configuration is resilient against failure; should a rotor or motor fail, it could still fly safely. In fact, flying would still be feasible if as many as four rotors were to fail, as long

RIGHT An early concept of Dragonfly, with the Mars Helicopter Scout shown to scale. The blunt curved nose of Dragonfly allows it to fit inside the aeroshell. *(APL)*

CENTRE Planetary exploration poses unique packaging challenges – the rotorcraft must fit inside an aeroshell for the hypersonic entry into Titan's atmosphere, prior to emerging to fly in its intended environment. The door in the backshell is to permit installation of the MMRTG on the launch pad. A cruise stage must provide propulsion, guidance and telecommunications during the long journey through space. *(APL)*

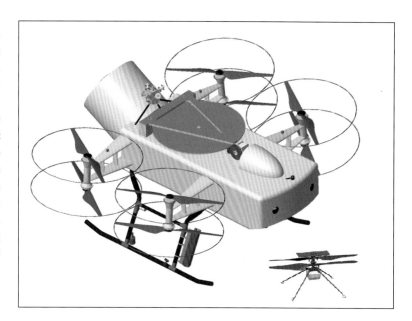

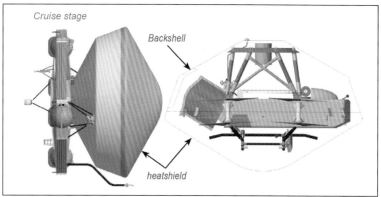

as two on the same corner continue working. This is important because there is no means of repairing or replacing a motor on a vehicle that is a billion miles away!

Given that so little sunlight reaches the surface of Titan, solar power is impractical – it will be necessary to use a radioisotope power source. After several years in space an MMRTG would put out less than 100 watts of electricity, but it would also produce about 2kW of heat, which would help warm Dragonfly's interior against the dense, deeply cold Titan atmosphere. This would be transferred the same way that heat is extracted from a car engine, using a pumped fluid loop. These pumps have proved themselves in space in particularly demanding thermal challenges, such as on Curiosity and the Parker Solar Probe.

During the night, Dragonfly will operate in a low-power mode, mostly just 'listening' with its seismometer and recording periodic weather readings while recharging its battery in time for sunrise.

As there will be no relay satellite, Dragonfly must communicate with Earth directly using a circular antenna deployed on its top deck. Panoramic cameras are piggybacked on the antenna to provide a high vantage point and eliminate the need for a separate pointing mechanism.

Titan's nitrogen-methane atmosphere at low temperature is a little less viscous than Earth's, so sound travels a bit more slowly. These factors, and the density, determine the

BELOW Two frames from an animation (dragonfly.jhuapl.edu) showing the arrival of Dragonfly. On the left, after about an hour on a small drogue parachute, a large main chute is deployed to slow the craft's descent and allow the front heat shield to fall away. On the right, the lander is released from the backshell and then commences the transition to powered flight by swooping down and ahead. The landing skids swing out prior to release, in order to allow the radar altimeter to lock onto the ground. *(APL)*

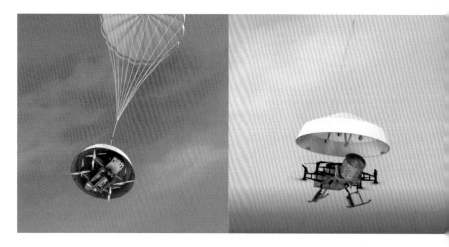

ABOVE **Dragonfly on the surface. The 1m high-gain antenna beams data directly back to Earth. This is flat and uses an array of slots to generate a narrow beam (rather like the antenna used by Japan's Akatsuki mission at Venus) and allows the antenna to pack-flat for streamlined atmospheric flight. The canted cylinder at the rear is the MMRTG that provides power and heat. After a local day or so at each location, a set of eight rotors powered by a large battery (charged up overnight) will move the vehicle to a new location.** *(APL)*

BELOW **A view of Dragonfly with antenna deployed, with various components labelled.** *(APL)*

best rotor shape and speed (for example, rotors become inefficient if their tip speed approaches the local speed of sound). In fact, the best blades for Titan (the low viscosity means that surfaces operate at a higher Reynolds Number than Earth) are shaped rather like those for wind turbines on Earth;

the chosen wing section does not rely on particular surface smoothness but is tolerant of dings or dirt. The low temperatures on Titan mean that the rotor motors must be warmed up before flight in order to soften the lubricants, just as in the case of the actuators on Curiosity.

As Titan is a billion miles away, light (and radio signals) take over an hour to cross space, so there is no hope of directly 'joysticking' a flight from Earth. It will require autonomous flight control and navigation. The basic autopilot will use an inertial measurement unit (IMU) with super-precise laser gyros and sensitive accelerometers. This is accurate enough for short flights, but cameras will be used to measure the flight speed over the ground during longer excursions. For even higher accuracy, Dragonfly's cameras will match up landmarks identified on previous flights. (There is no GPS on Titan, and Titan does not have a magnetic field, so compasses cannot be used.) This type of optical terrain relative navigation (TRN) has been recently developed for Mars and Moon landers, as well as terrestrial applications (APL has worked on digital scene matching for cruise missile guidance for decades). As shown by the Huygens probe,

BELOW **The Principal Investigator of Dragonfly with a full-scale prototype of one of its rotors, milled in aluminium (the real deal will be titanium, or maybe a composite with titanium leading edges, depending on weight, cryogenic tolerance and impact resistance needs).** *(Author)*

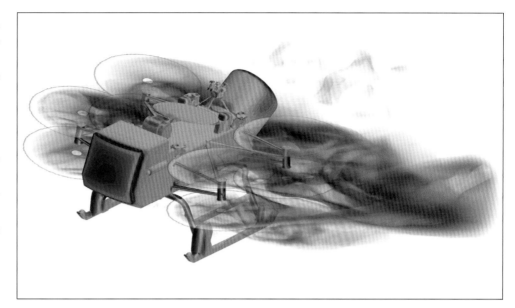

RIGHT The aerodynamic performance of Dragonfly will be confirmed in wind tunnel tests, but initial studies have used Computational Fluid Dynamics simulations. The blue traces show the simulated swirling airflow from the rotors, while the colours on the body show surface pressure: the purple and red show that the blunt nose and the MMRTG cause significant drag – future design work may streamline these areas. *(Mike Kinzel)*

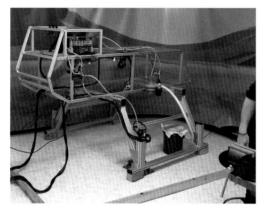

ABOVE The full-scale frame model of Dragonfly (the 'Iron Bird') used to test an all-up prototype of its sampling system, with drill, carousel, diverter valves, and blowers. *(Author, at Honeybee Robotics)*

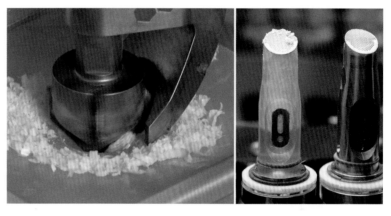

ABOVE A close-up of the conical drill for Dragonfly (left). The duct at the right vacuums up the cuttings through the pipes into one of the pen-sized cups of the sample-handling carousel (right). The oblong mesh window of the cups allows a laser to vaporise material for analysis in the mass spectrometer. *(Author/Honeybee Robotics)*

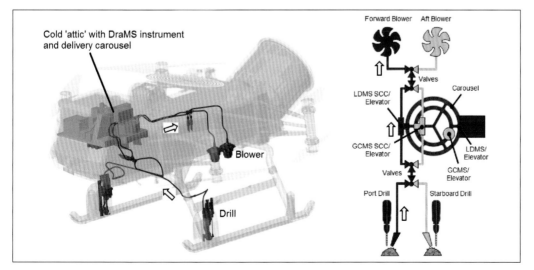

LEFT The flexible sampling system uses flow diverting valves to direct the airflow from one of two drills, through one of two parts of the instrument carousel, and out through one of two blowers. *(Author, at Honeybee Robotics)*

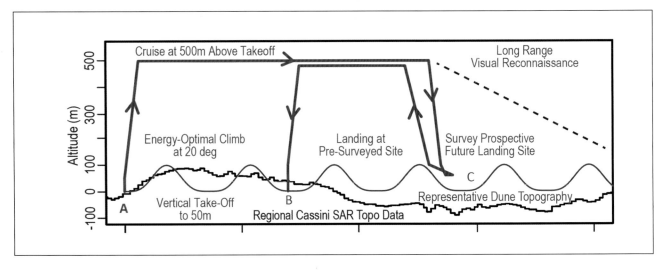

ABOVE Dragonfly's 'leapfrog' exploration strategy. The vehicle can fly autonomously, and make local real-time decisions for landing at a safe spot. But in general, the flight range allows potential future landing sites to be scouted during low-altitude reconnaissance flights whose data can be analysed on Earth prior to committing to a future landing. *(Author)*

there is sufficient light at the surface of Titan for imaging – indeed, the visibility was several tens of kilometres.

A 'flash LiDAR' will measure the roughness of the terrain immediately below the vehicle and enable Dragonfly to find a safe spot to touch down by itself, within a target zone specified by ground controllers in advance. Apart from the landing upon arrival, Dragonfly will typically only be ordered to a 'safe' site that it has already scouted from the air. In fact, for a scientific traverse involving multiple sites of various geological types, the plan is to 'leapfrog' to ensure optimum terrain for landing. On departing site 'A', the vehicle would scout region 'B', then return to 'A' which is known to be safe. Then it would take off from 'A', scout region 'C', and land at 'B'. This two-steps-forward-one-step-back strategy avoids the risk of heading into the unknown.

After Dragonfly enters Titan's atmosphere and flies out from beneath its parachute-suspended heat shield, the first landing is to take place over a dunefield where the interdune areas (2–3km wide) are typically very flat, often gravelly, plains. So even if Dragonfly emerges from its heat shield over a dune, it would need to fly only a couple of kilometres to find a flat spot on which it could safely land.

Once safely down, Dragonfly will point its antenna at Earth, begin to transmit data and report on conditions for a few days before the Sun and the Earth set. Titan's day (or 'Tsol') is 16 Earth days long, so Titan's night, when the lander is snoozing in its low-power mode and

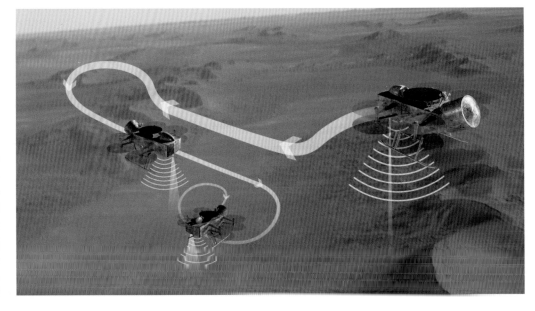

RIGHT With its multi-kilometre flying range, Dragonfly will be able to traverse dunes and interdunes to find a safe landing site using its on-board sensors such as a 'flash LiDAR'. *(APL)*

recharging its battery, lasts a week. This will give scientists time to examine the first data and refine plans for the next Tsol. Unlike the Mars rovers, where operators suffer 'jet lag' by having to work on Martian time, Dragonfly operations can be comfortably planned (for the most part) on regular office hours!

Although Dragonfly is expected to yield exciting findings in a whole range of scientific areas, to say nothing of spectacular pictures, the main motivation of the mission is to understand Titan's surface chemistry. We know there are abundant carbon-rich materials there, and there are also places where water has interacted with this material. When these kinds of reactions have been done in the laboratory in a few hours, many of the building blocks of life (such as amino acids, which make up proteins in living things, or pyrimidines, a molecule that stores information in DNA) can be created. But we don't know, over thousands or millions of years, how far along the chain of chemical complexity these reactions travel toward the functions like information storage or energy generation in a planetary environment like Titan. We think these chemical processes happened on Earth, but evidence of the details is lost. Titan is like a giant prebiotic chemistry laboratory where these crucial steps are being kept in deep freeze, waiting for us to pick them up.

Although Dragonfly does not have a robotic arm (that on the Curiosity rover is complex and heavy) it can interact with its environment in a number of ways. Its drills can acquire samples using rotary-only or rotary-percussive action. The texture of the cuttings can be examined by the cameras before deciding to ingest a sample. The percussion ('hammering') of a drill can be used to impart vibrations that the seismic sensor will 'listen' for, to deduce the stiffness of the ground. The sound of drilling will also be recorded by a microphone. A rotor can be spun up to various speeds to impose different levels of downward wind stress, and by observing with the cameras and electric field sensors it will be possible to determine the wind that is required to make sand move. This 'threshold wind speed' is an important parameter for decoding what the giant sand dunes mapped by Cassini have to say about Titan's climate history.

DRAGONFLY'S INSTRUMENTS

Instrument and Lead	Technique
DrACO (Drill for the Acquisition of Complex Organics) Honeybee Robotics	Dual rotary-percussive drills with custom bit. Pneumatic sample transfer system for rapid, cold transfer of drill cuttings to DraMS sample carousel.
DraMS (Dragonfly Mass Spectrometer) NASA Goddard Space Flight Center	Mass spectrometer with laser desorption and thermal/gas chromatography front ends to analyse molecular composition of ices and organic compounds in sampled material.
DraGNS (Dragonfly Gamma-Ray and Neutron Spectrometer) Johns Hopkins APL	Pulsed neutron generator, with gamma-ray and neutron spectrometers to quickly measure elemental composition (carbon, nitrogen, hydrogen, oxygen, salts etc.) of surface under lander.
DragonCam (Dragonfly Camera Suite) Malin Space Science Systems	Forward and down-looking wide-angle cameras. Pointable cameras for panoramas; close-up imagers to view drill sites and surface material. LED illuminators for colour information and organic fluorescence detection.
DraGMet (Dragonfly Geophysics and Meteorology Package) Johns Hopkins APL	Wind, pressure, temperature measurement and sensors for methane humidity and hydrogen gas. Electric field and physical properties of surface. Seismometer.

The Dragonfly sampling system has a drill on each of the two skids. The drill mechanism itself is evolved from that used by the Apollo astronauts on the Moon, with a specially designed conical drill bit for sampling Titan's (softer) materials. The drill cuttings will be transferred to the chemical analysis instrument through pipes by suction, much like a vacuum cleaner.

A mass spectrometer will measure the size of molecules either zapped from a sample using a laser or cooked off in a tiny oven. This is quite similar to the corresponding instrument on the Curiosity rover which recently detected organic compounds on Mars.

Scientists will help decide whether to take a sample at a given landing site using information from a gamma-ray spectrometer that 'illuminates' the ground with pulses of neutrons from a generator (a bit like an X-ray tube; again, the Curiosity rover has such a neutron source), then measures the neutrons and gamma rays that are emitted. The result will indicate whether we are on a deposit of frozen water and whether that ice is salty or not, or whether we are instead on organic-rich material. One helpful feature of the Titan environment is that the special high-purity germanium gamma-ray detector needs to operate below about 100K. Normally on space missions such detectors need mechanical cryocoolers (essentially little refrigerators) to get that cold, but on Titan we will be able to just hang the detector out in the cold breeze!

And of course, like any lander or rover, Dragonfly will have a suite of cameras. Down-looking wide-angle cameras on the vehicle's belly will provide mapping while in-flight and inspect the area under the lander to decide on sampling. Its close-up imagers will be able to see individual sand grains, and look at the sites where each drill will touch the ground. Although Titan's haze bathes the dayside in dim red light, Dragonfly will be able to take colour close-up views at night using a multicolour 'flashlight' employing light-emitting diodes (LEDs). One interesting feature of this is that Dragonfly will have ultraviolet LEDs, because many organic compounds that we expect on Titan fluoresce, glowing blue or green when illuminated by ultraviolet. You can try this for yourself by using a 'black light' to illuminate tonic water (which contains quinine) or laundry detergent (the 'brightening agent' is another fluorescent organic material). Dragonfly will also include forward-looking cameras for scouting from the air, and cameras mounted at the top of the communications antenna to make panoramic views of each landing site.

Finally, the instrument that I lead is a meteorology and geophysics package, using a suite of simple sensors. As well as air temperature, pressure and methane 'humidity', it will measure the wind speed with 'hot film' anemometers (following a design for the Beagle 2 Mars lander) where a small cylinder is heated and then the 'wind chill' in different directions is measured. Because the structure of a lander can distort the airflow slightly, there will be a wind sensor on each of the four rotor hubs so that at least two are 'upwind' of the vehicle and in undisturbed air. The wind sensors will also be used to check that conditions are safe for take-off. A seismometer can be lowered to the ground to listen for Titanquakes and perhaps diagnose how thick Titan's ice crust is. The skids also include measurements of the electrical capacitance beneath the vehicle skids, which will give the dielectric constant of the surface, and thermal measurements to assess whether the ground is damp.

In June 2019 NASA announced the choice of Dragonfly as the next mission in its New Frontiers program. It is to launch in 2026 and arrive in 2034, almost exactly one Titan year after Huygens. Because it will be targeted at a similar latitude, the Huygens data will be directly applicable for mission planning.

After Cassini and before Dragonfly

While Dragonfly is under development and in transit to Titan, we will have to content ourselves with telescopic observations. Ground-based telescopes (especially, but not only, large telescopes with adaptive optics systems) can monitor Titan in the near-infrared for cloud activity, to better guide models of the weather and hydrological cycle. Visible photometry (continuing the record from Lowell Observatory 1970–2013) will document year-to-year changes in the haze. Perhaps a snapshot or two of Titan in the visible from the Hubble Space Telescope will add some detail to this picture, as long as that telescope – now in space for over a Titan year – continues to operate.

When Hubble's successor, the James Webb Space Telescope (JWST) is finally launched, it will provide the opportunity to observe other changes in the atmosphere. In fact, Titan is too bright for some of its instruments to observe effectively, but it can make disc-integrated observations in the near-infrared with much better spectral resolution than could VIMS on Cassini and can monitor seasonal changes in the CIRS-like mid-infrared, especially at wavelengths where the Earth's atmosphere impedes terrestrial telescope observations.

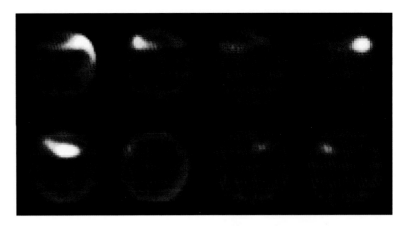

RIGHT Adaptive-optics near-infrared images of Titan acquired by the Gemini telescope in 2019, showing prominent cloud activity around the north pole. Sadly Cassini was no longer present to observe them more closely.
(Gemini Observatory/Alex Hayes/Cornell)

A particularly powerful tool to study Titan's upper atmosphere has come on line in recent years: the Atacama Large Millimetre Array (ALMA). Nitrile molecules like hydrogen cyanide (HCN), acrylonitrile (CH_2CHCN), hydrogen isocyanide (HNC) and so on, have characteristic line emissions in the millimetre range which can be measured rather precisely using ALMA's radio receivers. The frequencies of the lines are known exactly, and the strength and shape of a line are sensitive to the abundance profile of the molecule and the temperature structure. So if the Doppler shift on the line can be measured, ALMA's many, widely spaced, dishes can attain a spatial resolution sufficient to plot out the Doppler shift on the sky, thereby mapping Titan's changing winds with latitude.

So we see that telescopic observations will supplement ever-improving computer models of atmospheric circulation and geological evolution, and laboratory work on Titan's chemistry. The development of the Dragonfly mission will also motivate further analysis of the treasure trove of Cassini data. These diverse threads of science will weave a deeper understanding of this most remarkable world. Dragonfly's exploration of the Selk crater and its environs is rich with scientific promise,

and will be a thrilling journey for the public at large, but it must be remembered that even with its transformational mobility, Dragonfly will only explore a tiny fraction of the surface. We will have to await follow-on missions for a comparable global view of Titan, and in-situ study of its seas. The story related in this book marks only the beginning...

ABOVE The mammoth 6.5m primary segmented mirror of the James Webb Space Telescope during deployment tests. In the 2020s, JWST may provide important continued observations of Titan. *(NASA)*

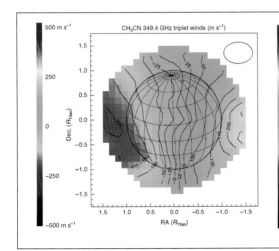
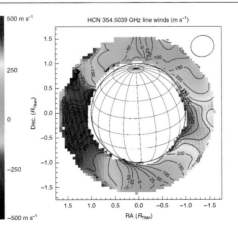
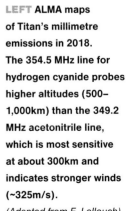

LEFT ALMA maps of Titan's millimetre emissions in 2018. The 354.5 MHz line for hydrogen cyanide probes higher altitudes (500–1,000km) than the 349.2 MHz acetonitrile line, which is most sensitive at about 300km and indicates stronger winds (~325m/s).
(Adapted from E. Lellouch)

FUTURE EXPLORATION

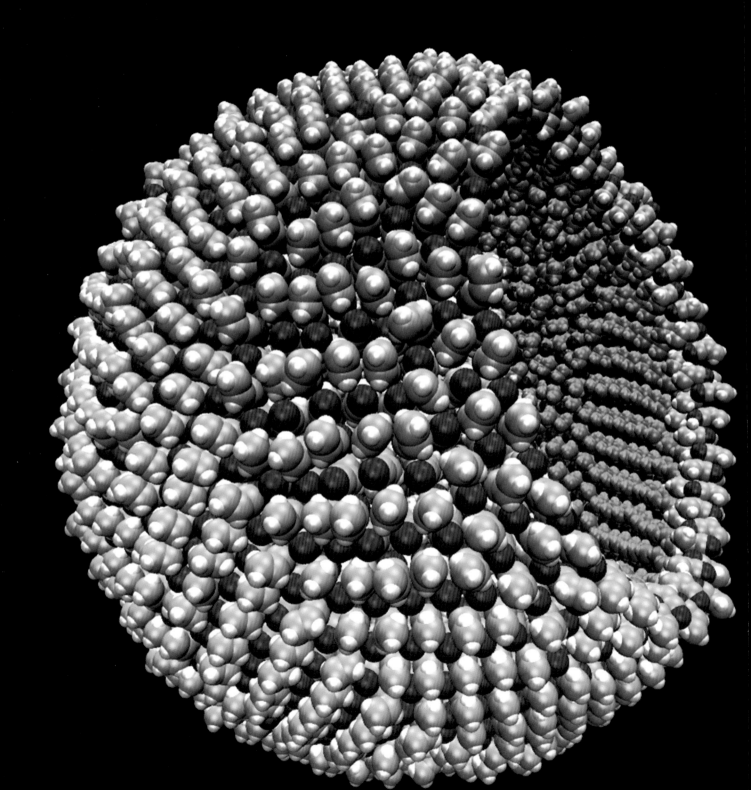

Chapter Ten

Life on Titan?

Titan's organic-drenched surface and liquid water interior give it key ingredients for life. Could life as we know it have evolved there? Could an exotic biochemistry function in a methane sea? The stuff of science fiction, perhaps.

OPPOSITE The computed structure of an 'azotosome' – a membrane using acetonitrile that could form a globule skin or vesicle in Titan's hydrocarbon seas in the same way that lipids (fats) can do in water in terrestrial biochemistry.
(James Stevenson/Ella & Alexander Tokarev)

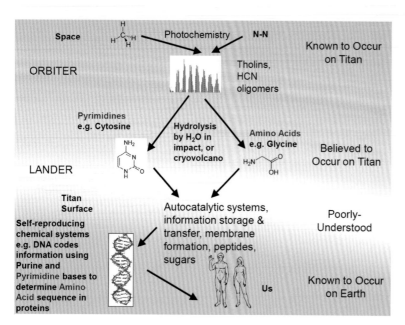

ABOVE **A schematic of the steps of chemical complexity, and how they map onto Titan exploration. We are fairly sure the building blocks are present in abundance, but how far have they assembled themselves?** *(Author)*

Part of Titan's fascination is its possible bearing on the origin(s) of life. In some respects Titan today (and more so, Titan in the past) resembles the early Earth. Before geochemical changes and, later, the origin of photosynthesis made our atmosphere oxygen-rich, it was once much more 'reducing', likely had more methane and perhaps traces of hydrogen. Since the Sun was ~25% fainter 4 billion years ago than it is today, the greenhouse warming from that reducing atmosphere may have prevented the planet from freezing. However – like Titan today – solar ultraviolet light would have attacked the carbon-hydrogen bonds in methane, making a range of organic compounds that wouldn't quite have been the same as on Titan today. A big part of Titan's chemistry is nitrogen, and ultraviolet light isn't energetic enough to break the nitrogen-nitrogen bond – on Titan this happens because the atmosphere is lanced by energetic electrons accelerated in Saturn's magnetic field. Whereas Titan's frigid atmosphere is virtually bereft of oxygen compounds, there would have been appreciable amounts of carbon dioxide and water vapour in the early Earth's atmosphere, hence the compounds making up the haze might have included more oxygen atoms. The balance of carbon, nitrogen and oxygen may well have been rather different in primordial Earth's haze than it is on Titan, although the extent to which this matters for both climate effects (Earth's haze may have been more reflective) and for the origin of life, isn't known. It is worth remembering that early Titan may have been warmer and more oxygen-rich due to the heat of its formation, and so early Titan and early Earth may have been much more similar than the other combinations.

Photochemical haze may give life a 'head start' by building up a variety of organic molecules. Other possibilities include cometary or meteoritic material – some meteorites, particularly the carbonaceous chondrites, have lots of carbon-bearing material, including amino acids. These may have been made by aqueous chemistry (in hydrothermal systems, for example) or by the action of cosmic rays or ultraviolet on ices like methane.

But the photochemical material on Titan is largely a dead end. The most functionally useful organic molecules require oxygen – for example sugars contain as many oxygens as carbons, amino acids that build proteins each have an oxygen atom, and so on. A trace of oxygen may be incorporated from the wisps of material from Enceladus that waft onto Titan's upper atmosphere, but this is very little. The most likely way to build large amounts of biologically relevant material, and more importantly for that material to advance beyond simple oxygen-bearing compounds, is for liquid water to interact with the photochemical organics.

There are three principal ways this could happen. Firstly, if surface material can be somehow drawn down into the interior water ocean. We have no evidence of anything resembling plate tectonics on Titan, and any sort of subduction process has something of a buoyancy challenge because, unless particularly exotic compositions are invoked, the ice crust will simply float on the ocean and there will be little or no force to drive surface material down.

The second, more promising mechanism is cryovolcanism. Although this too has a buoyancy challenge because water, and even water with appreciable amounts of ammonia in it, is denser than the ice crust, and so wouldn't buoyantly rise. However, if there are lots of gas bubbles in the water magma, perhaps hydrogen and/or methane from the serpentinisation of rocks in the deep interior, then a sufficiently thick cryolava flow could persist in a liquid state to react with any tholin or other organics that it overran, for centuries or longer.

The third process is, of course, impact. When a large comet or asteroid forms an impact crater, some of the target material is melted or even vaporised, and the fraction of the crater volume transformed this way increases with the impact energy. Indeed, one of Earth's largest nickel deposits, at Sudbury, Canada, formed this way – a large sheet of molten rock formed, and the dense metal-bearing sulphide minerals sank to the bottom of the molten layer. The hydrolysis reactions that turn tholins into pyrimidines, amino acids and other prebiotic compounds are of course temperature-dependent. The production of these compounds may proceed millions of times faster in an impact melt sheet at 273K (or a higher temperature) than in a cryovolcano at the lowest melting temperature for water-ammonia of 176K. This was one reason for selecting an impact crater (Selk) and its ejecta blanket as the target region for exploration by Dragonfly.

Since this sort of material can make available chemical energy, it can serve as food. Indeed, the experiment has been done – cultures of bacteria have been seen to grow and flourish on tholin in the laboratory; literally manna from heaven on Titan.

But there is of course a big difference between an environment that can support living things, and one that supports the origin of life, that magical transition from non-living to living. We don't know how that transition occurred on the early Earth (or Mars), only that it must have happened at least once somewhere.

Consider that iconic molecule DNA (deoxyribose nucleic acid), the molecule that encodes the 'blueprints' from which many living things are constructed. The 'alphabet' that specifies the sequence of amino acids needed to assemble the proteins that make up the functional parts of living matter has four letters – A, C, G, and T – corresponding to the pyrimidine bases thymine and cytosine, and the purine bases adenine and guanine. DNA can replicate the information because A always matches up to T, and G to C, so an 'unzipped' half of a DNA sequence can serve as the template to rebuild the other half.

The pyrimidine bases, a ring of carbons with several nitrogens, can be readily made by tholin hydrolysis. The purine bases have two rings, so are a little more complicated, but perhaps not impossible. However, what about

LEFT In this view (ISS basemap with VIMS false colour) of the Sinlap impact structure, blue signifies water ice. Impacts may have both excavated and heated water-rich material, creating a natural cauldron of prebiotic soup! *(S. Le Mouelic/VIMS Team)*

BELOW The structure of DNA. At least some of the components (thymine and cytosine) are easy to make on Titan, but others may not be. *(CC0: Madeleine Price Ball)*

the 'backbone' on which the alphabet sits? One can imagine the sugar (ribose) bit being made by hydrolysis, but the phosphorus in the molecule might be an interesting bottleneck – most phosphorus in nature is not chemically accessible. In fact, the late availability of phosphorus from meteorites may have been crucial to the origin of life on Earth as the original inventory may have been sequestered away in insoluble rocks.

In the case of the former, the internal oceans of icy satellites like Titan might well constitute a 'habitable' environment in so far as some terrestrial biota might survive there, or perhaps not. Titan's internal ocean may be rich in ammonia, which some biota may not tolerate, but others enjoy. (The same is true for the other icy satellites – for example, there could be large amounts of sulphuric acid present on Europa.)

A crucial question is whether any exchange of material between Titan's deep interior and the surface is possible. Getting internal ocean liquid onto the surface will suffer the challenge that water is more dense than ice. But with enough ammonia in the liquid, this negative buoyancy problem is greatly reduced – ammonia is also the most powerful antifreeze, allowing liquid to reach the surface more easily without freezing. And only a little gas would be needed to perk up the ascent rate of liquid through the ice crust: perhaps Doom Mons and other such features show that it already has. And the evidence of Ar^{40} in the atmosphere shows that at least *some* gas has made it from the potassium-bearing silicate core to the surface. So perhaps if there are bugs in Titan's ocean and they are delivered to the surface, we will be able to investigate their frozen remains.

Despite the immense distances between the planets, small amounts of material are exchanged between them. In an impact, small amounts of target material are spalled off at hypervelocity speeds. The existence of meteorites on Earth known to have originated from the planet Mars is tangible evidence for this process. It has been calculated that hundreds of tons of material may have been launched from Earth and arrived at Titan. This would have occurred particularly in the early history of the solar system, when bombardment was still heavy as the protoplanetary disc was slowly being cleared up. If this happened early enough in Titan's history, the ice crust would have been thin (or even non-existent) and material deposited on the surface could more easily be introduced into the internal water ocean. If that happened, and if the meteorites were carrying early terrestrial lifeforms, then that biota, or its chemical signature, may be present in the ocean today.

How would we know? There is a property of some organic compounds wherein they can be structurally identical except for being mirror images of each other, and living things tend to be selective in terms of chirality. Among such compounds are sugars and some amino acids. The property of having these mirrored structures is called stereoisomerism, and was discovered by the influence of these

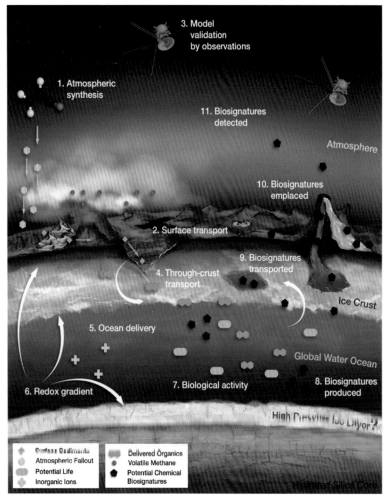

BELOW A schematic of the possible exchange of biosignatures between Titan's interior and surface. *(Athanasios Karagiotas/Theoni Shalamberidze/NASA/JPL/NAI)*

compounds on the transmission of polarised light – their crystals rotated the light plane to the left (L-) or the right (D-). It turns out that all the stereoisomer amino acids used by life on Earth use the L-form, and in fact if we eat the D-form, we cannot process it. The same is true of sugars, except in that case it happens that we use the D-form. The selectivity of stereoisomers can be a life-or-death issue in some drug designs.

There isn't an overwhelming reason why life should prefer L-amino acids and D-sugars, any more than it is impossible for cars to drive on the right-hand side of the road, say. Obviously once a convention is established, everyone must follow the same one – unless a contrary sub-population can be isolated from the rest, by the seas for example, so that left-hand drivers are not thrown into opposing traffic. And so it may be with stereoisomerism – life on Titan, if it exists, will probably have a preference. If it happened to be for L-amino acids, that wouldn't prove anything – the coin could have been flipped on Titan with the same result as Earth, or possibly Titan and terrestrial life had a common origin – Mars, perhaps! But if life on Titan is found to have opted for D-amino acids, that would imply a distinct origin of life – a powerful conclusion!

Alien life chemistry

Might there be monsters in Titan's seas? Dolphins would love Titan – the low gravity would allow them to leap formidably. But Titan is too cold for life as we know it, and although there have been interesting studies of bacteria in oilfields and the Pitch Lake of Barbados, even these organisms rely on tiny little blobs of water among the hydrocarbons.

So the question we have to ask is whether there might be an alternative chemistry of life. There are chemical systems on Titan that might be able to support metabolism, or, in other words, provide free energy. One possibility is acetylene and hydrogen (which was discussed earlier) and there are surely others. But at low temperatures, would these react fast enough, and can the right chemistry operate in a solvent other than water – specifically a non-polar solvent such as ethane? (Methane, in fact, is not a very effective solvent for anything, so ethane-rich fluids are likely a better bet.)

As regards information storage, is there a structured polymer analogous to DNA? So far, we simply don't know. But the function of membranes, at least, does have a candidate molecule. Living things need boundaries to keep their finely tuned innards from diffusing into oblivion, and a compound that can form vesicles (i.e. the boundary of globules) in liquid hydrocarbons has been determined to exist in some abundance on Titan – acrylonitrile. Like the fatty acids which, in part, do this function for us, this molecule has a polar end (the -CN bit) and a non-polar end. Thus the molecules all align to put the non-polar end outward toward the ethane-rich solvent, and bottle up the more interesting polar molecules inside.

So we are aware of a few pieces that could work but we don't (yet) know about

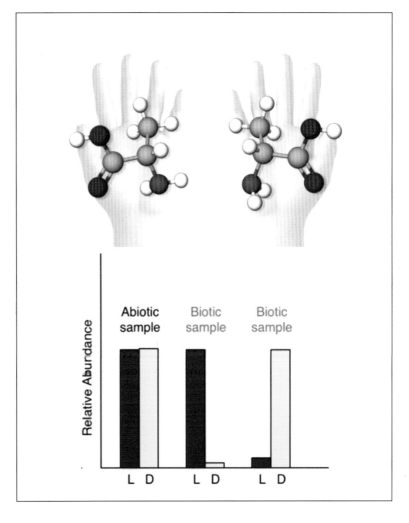

BELOW Some organic compounds like sugars and amino acids (in this case the amino acid alanine) can have mirror-image forms. Abiotic chemistry tends to produce both in equal amounts, but living things tend to exploit only one. *(NASA/JPL)*

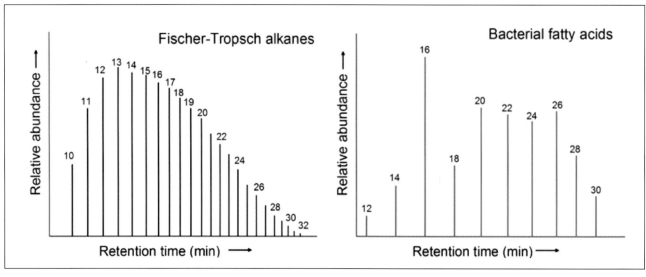

ABOVE **Because life is efficient, it exploits existing building blocks and so biological samples tend to have a selectivity (in this case it is even-numbered carbon atoms), whereas abiotic processes tend to produce a random suite of compounds.**
(NASA/JPL)

others. And of course we don't know how the ingredients might magically combine to yield self-propagating chemical systems and transcend the murky line between non-life and life. Significantly, we do not know where and how often (it was likely more than once) this transition occurred on Earth. Perhaps it was in a 'warm little pond' where the vital chemicals could be concentrated by tidal cycles and evaporation, but recirculating fluids in a 'black smoker' seafloor hydrothermal vent can't be ruled out, nor some other environment entirely. If we find chemical systems on Titan that have taken even a couple of steps on the road of increasing complexity, then we will learn much about life as a cosmic phenomenon.

How might we recognise a different chemistry of life? One thing to look for is patterns. Living things have been honed by evolution to be lazy (or at least, efficient). So rather than assemble a large functional molecule entirely from scratch, this is often built up from smaller elements that are common to other functional molecules – i.e. building blocks. This is sometimes termed the Lego Principle, and is seen for example in the distribution of carbon chain lengths in biological material such as fats. Whereas an abiotic synthesis such as the Fischer-Tropsch reaction would randomly make a bit of everything, with the probability of forming the larger molecules falling as the number of steps involved in making them progressively increases, fatty acids from living things have only even-numbered carbon chains because they are assembled using units of two carbons. By itself, finding such a pattern would not likely be a persuasive indication of life on Titan, but it might be an important clue.

Humans on Titan

Perhaps the most likely scenario for life on Titan is that one day humans will travel there. The journey would be long and arduous. But not only is Titan a fascinating world to explore scientifically, with recreational charms peculiar to its low-gravity, high-density environment, the abundance of readily accessible volatiles will surely make it an important outpost as a fuelling stop in any solar-system-wide civilisation, as Arthur C. Clarke recognised in his 1975 novel *Imperial Earth*.

It might be possible for Titan's methane to be readily exploited by processing the atmosphere directly, distilling it out of the air like beetles in Namibia that harvest dewdrops. Or perhaps landing in or close to Titan's seas would be worth the complications of operating

In certain ways, Titan is the most hospitable extraterrestrial world within our solar system for human colonization.

Robert Zubrin,
Entering Space: Creating a Spacefaring Civilization, 1999

in a marine environment – one could just splash down in Ligeia Mare and suck the nearly pure methane aboard through a hose.

Of course, there is no oxygen in which to burn the methane. One could perhaps melt some ice and electrolyse it, but for a civilisation that has already made it to Titan this would be clunky – one would surely use a compact nuclear reactor to heat methane or hydrogen and expel the hot expanding gas through a nozzle. A nuclear rocketplane would be just as effective for travelling long distances on Titan as it would be for flying to and from space.

However large-scale transportation is achieved, it is interesting to think about how we would get around locally. Bipedal walking is an odd means of locomotion. Essentially the body is an inverted pendulum, with its bulk swinging across the foot prior to swinging the other foot in front to replace it. It is, then, a process of continual falling, regulated by gravity. It changes if gravity changes. This was shown by the Apollo moonwalkers, whose most efficient means of locomotion was skipping or bounding (two-legged hopping).

Walking is most efficient when the leg swing takes place at the natural pendulum period – in normal Earth gravity this happens at a speed V=1.3m/s where the Froude Number is 0.25. The latter, named after a 19th-century naval architect – the number is also important for wave dynamics on the sea – is just V^2/gL, where g is the acceleration from gravity and L is the relevant length (here, that of the leg). It is basically impossible to walk at a Froude Number exceeding 1, as the foot needs to leave the ground, and in practice locomotion tends to switch to running when the Froude

ABOVE Although pressure suits are not required on Titan, bulky insulation would be, and a helmet with visor simplifies the management of oxygen and heat. Astronauts on Titan would enjoy the ability to jump in the Moon-like gravity. *(Walter Myers/Stocktrek Images/Alamy)*

LEFT Aided by a floodlamp in the dim twilight, astronauts deploy a weather balloon on Titan. *(Walter Myers/Stocktrek Images/Alamy)*

Number is ~0.5. On both Titan and the Moon, the optimum walking speed should be about 40% of that on Earth.

The situation for walking on Titan may be a little different from the Moon, in that the Apollo astronauts were in bulky suits which were not only heavy but also had an internal pressure of about 1/4 of an atmosphere. The rigidity of the suits meant that the astronauts had to make an extra effort to bend the knee joint, to the extent that they struggled to bend down to lift objects. But walking inside a pressurised habitat on the Moon will be pretty much the same as walking inside a habitat on Titan.

Interestingly, in principle you would not need a pressure suit outside a habitat on Titan. The atmospheric pressure is about the same as that at the deep end of a swimming pool on Earth. You would just need an oxygen supply, an insulated suit, and goggles or a face mask to keep the cold air from the eyes.

The insulation for Titan should be about three to four times better than that for the harshest terrestrial conditions (since the difference between the skin temperature and the air might be ~60°C for Antarctica and ~200°C on Titan). That would be rather bulky, maybe a thickness of 5–10cm. A more convenient alternative would probably be to have thinner insulation in some areas, and use electrical heaters to counter the 'leak' in those spots. A battery pack of a kilogram or two (which would itself need to be kept warm) would suffice for a few hours.

Any uncovered skin would become frostbitten if left exposed. As some Titan researchers know from experience in the lab with the boil-off from liquid nitrogen, brief unprotected operations are possible in air that cold – so long as one doesn't touch anything solid.

Titan's atmosphere is 'breathable' in the sense that human lungs could push air in and out, but one would (painlessly) lose consciousness in less than a minute due to oxygen deprivation. A simple heated oxygen mask would permit sustained activity in the Titan atmosphere. A few of the molecules present on Titan (like carbon monoxide and hydrogen cyanide) are outright toxic, but are present at concentrations too low to cause immediate harm. Many organic compounds are carcinogenic too, so long-term exposure on Titan may need some caution, or countermeasures such as drugs. Organic dust traipsed into a warm habitat would evaporate, so scrubbers would be needed to purge the air (interestingly, flora such as houseplants can be quite effective at removing traces of some organics). On the other hand, the radiation hazard from the trip through space to reach Titan might be more likely to cause problems than the materials present on its surface.

Although Titan's gravity is similar to our Moon's, it is not obvious that driving around would be a great means of exploring long distances on Titan. In fact, the Apollo lunar rover required considerable driver attention at speed because it tended to lose contact with the ground. In any case, there are sure to be places on Titan where the trafficability of the surface cannot be relied upon – muddy streambeds, and the loose slip-faces of sand dunes are just two examples of terrain hazards familiar on Earth where mobility could be impaired.

Rather better, surely, to fly. In fact, with an adequate wing area, sustained flight by a human should be possible under simple muscle power alone. An excellent challenge for aeronautical designers may be to determine the best such means of transport. In principle, one could flap large lightweight wings like an angel and take off on Titan, but a more practical configuration might be a powered hang-glider, with the operator in something that resembles a recumbent bicycle in order to use the power of leg muscles to drive a propeller.

More generally, heavier-than-air transport will be the best way to travel long distances, as on Earth. Lighter-than-air travel is possible, but likely to be unappealing for most users, because they will want to travel more rapidly than a few metres per second.

The first visitors to other worlds will of course live in the ships that brought them there. But beyond that, perhaps the first habitats on our Moon may be lava tubes – long tubular caves formed by volcanic eruptions. These are conveniently near the surface, with holes ('skylights') that can

ABOVE A prototype for a base on Titan? The German Neumayer station in Antarctica stands on stilts to remain above snowdrifts. Heat leaking into the ground and undermining a station would be a concern on Titan, so stilts that prevent thermal contact might make sense.
(Alfred-Wegener-Institut/Stefan Christmann CC-BY 4.0)

provide convenient access points. But their main advantages are an ambient structure with a few holes that can be plugged to form a pressurised environment, and thick walls that screen against radiation and micrometeoroid bombardment (these are not concerns on Titan).

A common trope in science fiction is a dome. Although this is a logical shape for a pressurised habitat on Mars or on some airless world it is not so straightforward for Titan, on which winds could ruffle a flexible membrane, and where the curved surface might act as a wing to loft the whole city like a bouncy castle in a windgust. Rather, a low-profile rigid structure is probably more practical, with only a slight positive pressure to exclude Titan contaminants. It will need to be insulated – not so much to preclude losing heat to the air, as to avoid heating the ground beneath and destabilising it. In that respect, Titan habitats may resemble some structures that are built on ice or permafrost, jacked up on legs or

> Someday, people will live on Titan, the largest moon of Saturn... They will go boating on lakes of liquid methane and fly like birds in the cold, dense atmosphere, with wings on their backs. This will happen because, at a certain point, it will make sense.
>
> **Charles Wohlforth and Amanda Hendrix**, *Beyond Earth*, 2016

RIGHT The changing orbit and polar alignment of Titan cause the distribution of sunlight at high latitudes to change (the same effect drives our Ice Ages). When humans and Neanderthals still coexisted on Earth, Titan's seas may have dominated its south, but today's 'hotter' southern summers have driven methane into the northern hemisphere, a pattern expected to continue into the future. *(Author)*

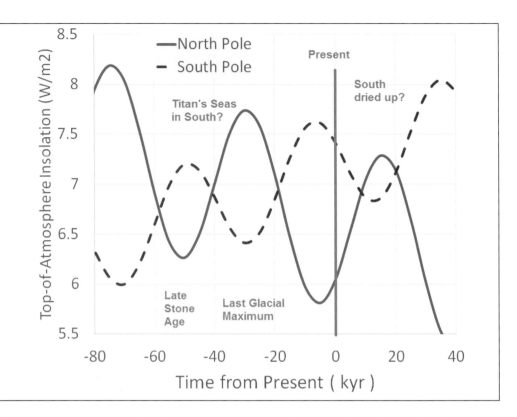

OPPOSITE Our Sun, shown here erupting a filament of the plasma that stimulates aurorae in planetary magnetospheres, will eventually become a red giant. As it expands and brightens, it will make Earth uninhabitable. At the same time, it may make Titan an environment suitable for a new origin of life. *(NASA/GSFC/SDO)*

stilts to keep the ground cold and to avoid the accumulation of wind-driven particulates on walls and doors. The structure will need firm attachment, perhaps with guy-wires, as the warm air inside will be buoyant.

The distant future

Whatever the state of biology or prebiotic chemistry is on Titan today, it is sure to get more interesting in the long run. Titan's climate will see astronomical (Croll-Milankovich) cycles, with the distribution of liquid and the wind patterns changing on timescales of tens to hundreds of thousands of years, but for the foreseeable human future, average temperatures on Titan will stay largely constant. On million-year timescales, the methane in the atmosphere may run out, reducing the greenhouse effect, or perhaps cryovolcanism will increase it. But one thing is for sure, on the longest timescale, over billions of years, all the climates in the solar system will get much warmer.

Our Sun is getting steadily brighter (by about 30% over the last 4 billion years) and models of the 'main sequence' evolution of stars suggest that in ~5 billion years the Sun will have brightened severalfold (in total luminosity), cooled (become redder) and expanded considerably to become a red giant.

Although the Earth may just escape being engulfed, it will be burnt to a crisp (and in fact will by then have become uninhabitable due to the loss of water through the stratosphere to space, much as methane is being lost from Titan today). However, ten times farther away, Titan may reach rather comfortable temperatures and the ice crust will melt.

There would likely be no dry land on this waterworld, although the deep global water-ocean may be covered by a slick of organics. Such an environment may be conducive to much more elaborate chemistry and may offer an entirely new opportunity for life to emerge.

In closing, we should note that while Titan is unique in our solar system today (recalling that perhaps early Ganymede, for example, was more like early Titan than it is today) there must surely be countless Titan-like worlds orbiting planets in other star systems. Since larger stars burn through their main sequence more rapidly, reaching the red giant state earlier, what we have just projected for the future of Titan will probably have played out already, somewhere out there...

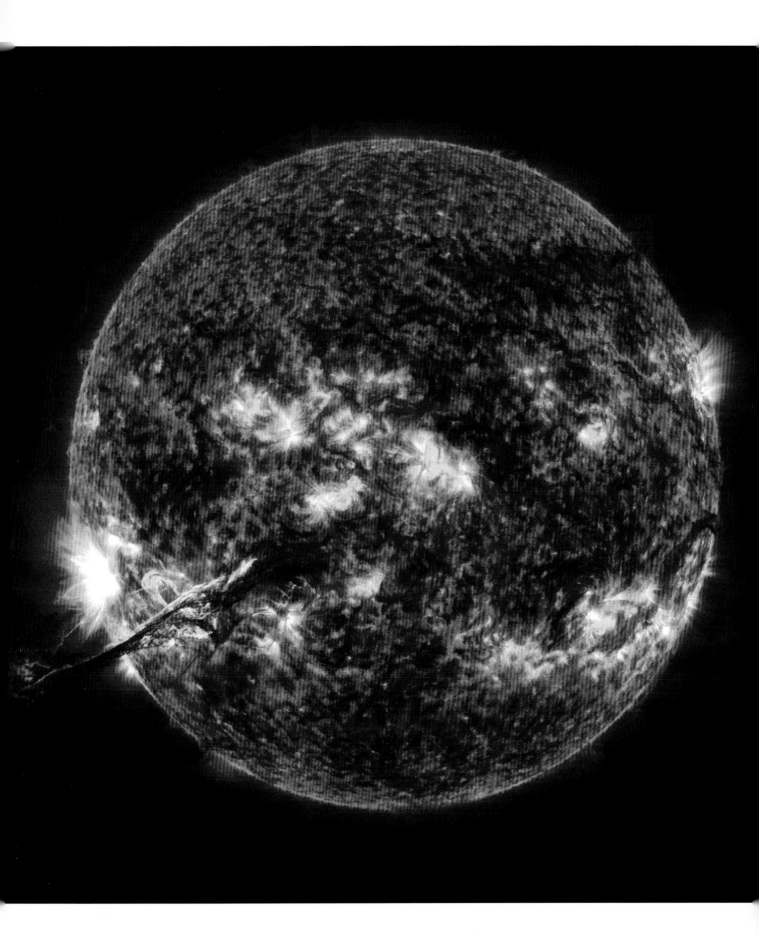

Maps

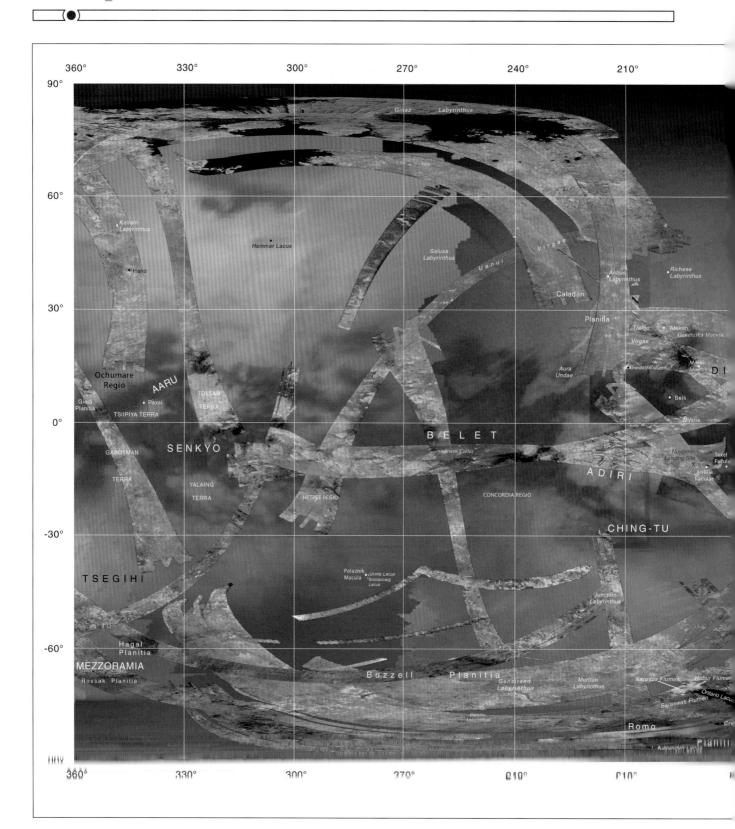

A VIMS basemap with radar swaths superposed and place names added. *(USGS)*

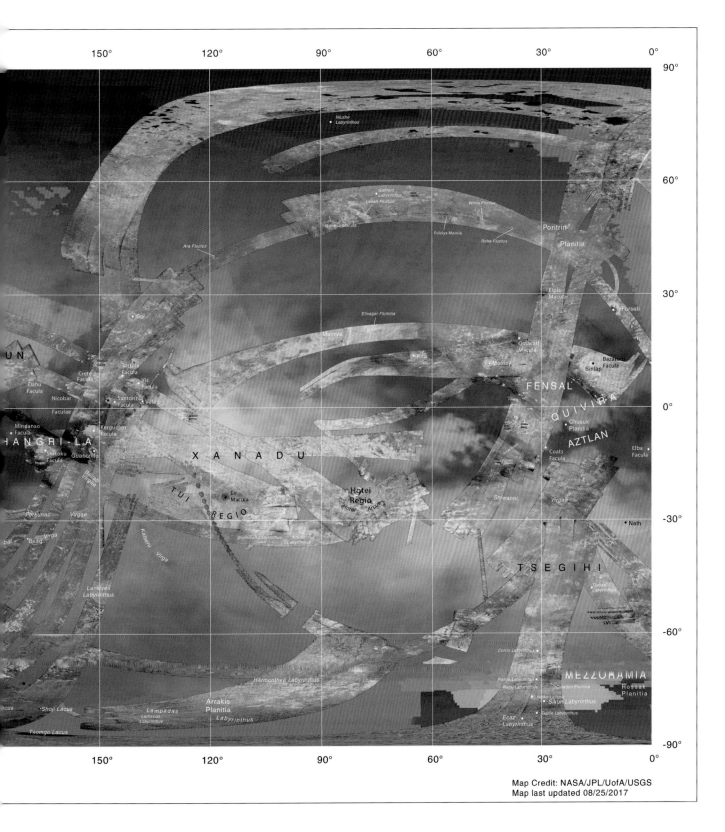

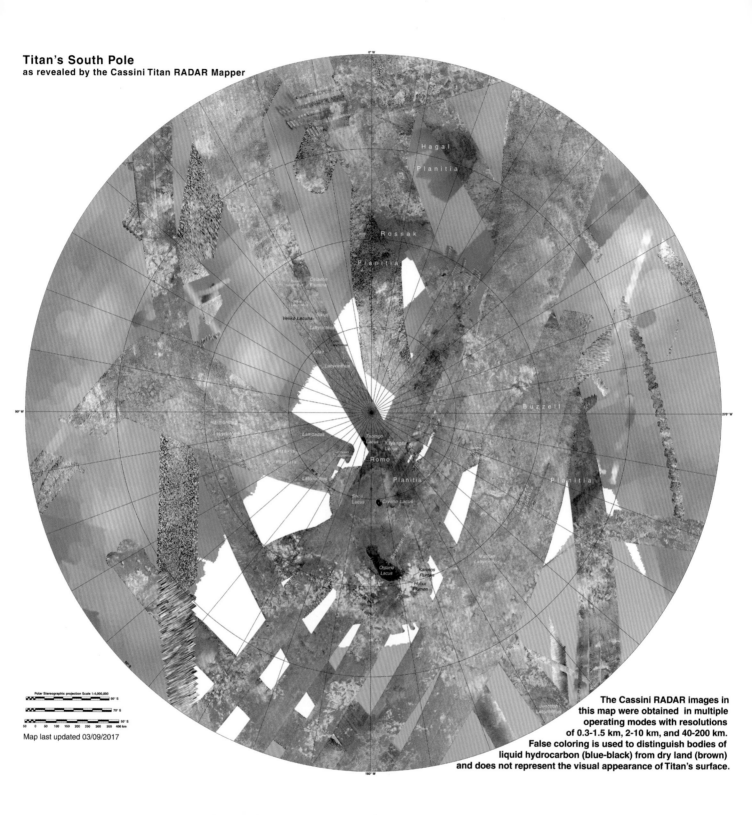

South polar radar mosaic with place names. The colour scale is interpreted... blue and black shading is low reflectivity characteristic of ... ..., higher reflectivity is in the gold colour. The paucity of liquid in the south is obvious. (USGS)

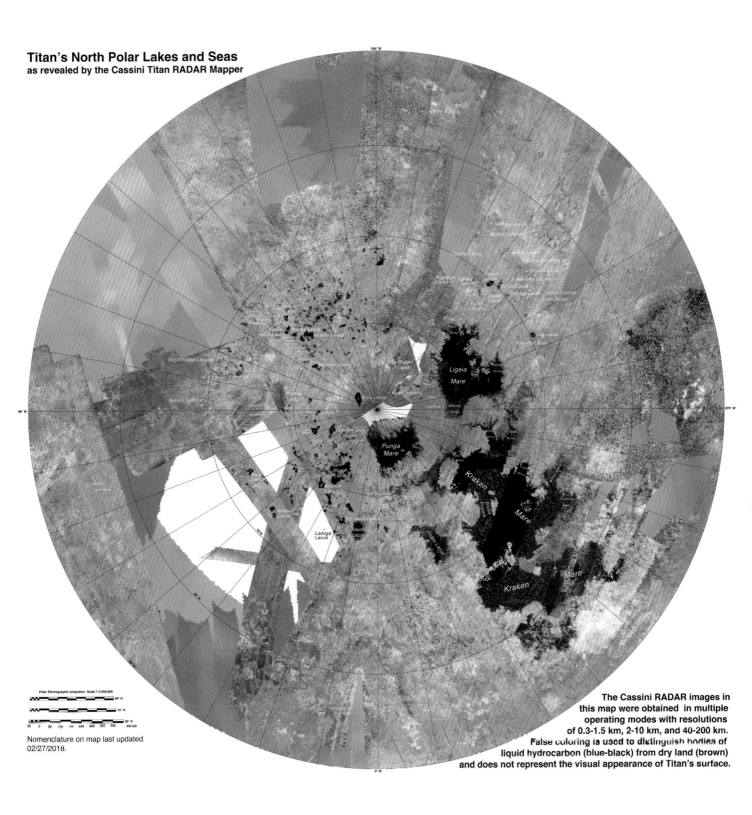

North polar radar mosaic with place names. The large number of small lakes is evident.

List of place names

Albedo Features	Dia (km)	Lat (deg)	Lon (deg W)	Name Origin, Feature remarks
Aaru	0	10	340	Egyptian abode of the blessed dead. Dark lane north of Senkyo.
Adiri	0	-10	210	Melanesian afterworld where life is easier than on Earth. Prominent bright region West of the Huygens landing site.
Aztlan	0	-10	20	Mythical land from which the Aztecs believed they migrated. Large dark lane, forms southern branch of the Fensal-Aztlan 'H' feature.
Belet	0	-5	255	Malay afterworld reached by a flower-lined bridge. Prominent dark region (informally 'the Heart of Darkness') visible in Hubble images; where Titan's dunes were first fully recognised on T8.
Ching tu	0	-30	205	Chinese Buddhist paradise where those who attain salvation will live in unalloyed happiness. Dark lane south of Adiri.
Dilmun	0	15	175	Sumerian garden of paradise, primeval land of bliss. Bright region north of Shangri-La, northeast of Selk crater.
Fensal	0	5	30	In Norse mythology, magnificent mansion of Frigga, to which she invited all virtuous married couples to enjoy each other's company forever. Prominent dark lane visible from Earth, forms northern branch of the 'H'; bounded on northeast by Sinlap.
Mezzoramia	0	-70	0	Oasis of happiness in the African desert, from an Italian legend. Dark area in the south with evidence of surface liquid.
Quivira	0	0	15	Legendary city in the American Southwest; site of a fabulous treasure sought by Coronado and other explorers. Bright lane that separates Fensal and Aztlan. Has Bohai Sinus at its western margin.
Senkyo	0	-5	320	Japanese ideal realm of aloofness and serenity, freedom from wordly cares and death. Dark area, broken up by bright streaks, a little east of the sub-Saturn point.
Shangri La	0	-10	165	Tibetan mythical land of eternal youth. (NB sometimes spelled Shangri-La, but the spelling in Hilton's Lost Horizon – without the hyphen – is the offical one.)
Tsegihi	0	-40	10	Navajo sacred place. Prominent bright region southeast of Xanadu.
Xanadu	3400	-15	100	An imaginary country in Coleridge's *Kubla Khan*.
Antilia Faculae	260	-11	187	Archipelago corresponding to the mythical island of Antilia, once thought to lie midway between Europe and the Americas. Small bright patchy area just to the east of the Huygens landing site.
Bazaruto Facula	215	11.6	16.1	Mozambican island.
Coats Facula	80	-11.1	29.2	Canadian island.
Crete Facula	680	9.4	150.1	Greek island.
Elba Facula	250	-10.8	1.2	Italian island.
Kerguelen Facula	135	-5.4	151	French subantarctic island.
Mindanao Facula	210	6.9	174.0	Philippine island.
Nicobar Faculae	575	0	159	Indian archipelago.
Oahu Facula	465	5	166.7	Hawaiian island.
Santorini Facula	140	2.4	145.6	Greek island also known as Thira.

Albedo Features	Dia (km)	Lat (deg)	Lon (deg W)	Name Origin, Feature remarks
Shikoku Facula	285	-10.4	164.1	Japanese island. Observed early in Cassini mission – superficial resemblance led to informal reference to it as 'Great Britain'.
Sotra Facula	235	-12.5	39.8	Norwegian island. More usually the suspected volcanic feature (Sotra Patera) is meant.
Texel Facula	190	-11.5	182.6	Dutch island.
Tortola Facula	65	8.8	143.1	Island in the British Virgin Islands. Site observed on first flyby by VIMS instrument. Speculative cryovolcanic interpretation not now widely held.
Vis Facula	215	7	138.4	Croatian island.
Eir Macula	145	-24	114.7	Norse goddess of healing and peace. Odd near-IR dark spot in Xanadu.
Elpis Macula	500	31.2	27	Greek goddess of happiness and hope.
Ganesa Macula	160	50	87.3	Hindu god of good fortune and wisdom. Observed on TA by radar, was suspected of being a volcanic dome, but subsequent observations did not support this interpretation.
Genetaska Macula	24	23.5	196.3	Peace Queen of the Iroquois.
Omacatl Macula	225	17.6	37.2	Aztec god of good cheer and lord of banquets.
Polaznik Macula	346.9	-41.1	280.4	Slavic god of New Year's happiness.
Polelya Macula	175	50	56	East Slavic god of matrimonial happiness.

Regions (Regiones), Lands (Terrae)				
Concordia Regio	1500	-20	241	Roman abstract divinity of harmony. Bright area at southern margin of Belet: site of surface darkening and rebrightening due to equinox rainstorms in 2009.
Hetpet Regio	1080	-22	292	Egyptian female personification of happiness.
Hotei Regio	500	-26	78	God of contentment, good fortune, and cheerfulness in Japanese Buddhism. Prominent 5-micron bright region, bounded by Hotei Arcus on the south. Initially speculated to be cryovolcanic, consensus is now that it is lakebed deposits.
Ochumare Regio	939	10.4	348.1	Puerto Rican goddess of happiness and weather. Bright region just north of Aaru.
Tui Regio	1200	-24.5	124.9	Chinese goddess of happiness, joy, and water. Prominent 5-micron bright area at southwest corner of Xanadu. Striking mountains and river channels.
Garotman Terra	970	-13.5	348	Iranian paradise where the souls of faithful men dwell. Bright region, looks like a bunny rabbit, west of Senkyo.
Tollan Terra	800	6.4	322.7	Aztec paradise where crops never fade. Dark region, broken up by many brighter streaks.
Tsiipiya Terra	573.24	2.83	340.12	Hopi name for Mount Taylor (New Mexico, US), as a sacred mountain. Bright lane between Senkyo and Aaru; has Paxsi ring at its edge.
Yalaing Terra	980	-19.5	324	Australian spirit land filled with game and clean water for good souls.

Plains (Planitiae), Dunes (Undae) and Streaks (Virgae)				
Arrakis Planitia	337.4	-78.4	117	Planet from the Dune series; home of the Fremen, a group of fictional people whose culture revolves around preservation and conservation of precious water.
Buzzell Planitia	870	-66.3	262.7	Planet from the Dune series; a cold planet, known for its lustrous Soostones, used as a 'punishment planet' by the Bene Gesserit.

Albedo Features	Dia (km)	Lat (deg)	Lon (deg W)	Name Origin, Feature remarks
Caladan Planitia	2800	31	226	Planet from the Dune series; the original home of the House Atriedes, a fictional noble family.
Chusuk Planitia	125	-5	23.5	Planet from the Dune series, known for its musical instruments.
Giedi Planitia	303.26	5.22	357.02	Planet from the Dune series; Giedi Prime is the homeworld of House Harkonnen.
Hagal Planitia	435	-60.59	344.95	Planet from the Dune series; a jewel planet known for its quality quartz.
Poritrin Planitia	1900	48	24	Planet from the Dune series; third planet of Epsilon Alangue, considered by many Zensunni Wanderers as their planet of origin.
Romo Planitia	400	-82.79	200.96	Planet from the Dune series; the planet on which the younger brother of Miles Teg was poisoned.
Rossak Planitia	512	-71.04	355	Planet from the Dune series; the source planet for the Bene Gesserit's original poison drug, or, the 'Water of Life', and the origin of the Bene Gesserit order itself.
Aura Undae	490	13.79	226.86	Greek Titanis goddess of the morning wind.
Boreas Undae	260	-6	215	Greek god of the north wind.
Eurus Undae	220	-7.5	210.3	Greek personification of the east wind.
Notus Undae	530	-10	211.1	Greek god of the south or southwest wind.
Zephyrus Undae	130	-8.5	217.1	Greek personification of the gentle west wind.
Bacab Virgae	485	-19	151	Mayan rain god.
Hobal Virga	1075	-35	166	Arabian rain god.
Kalseru Virga	630	-36	137	NW Australian rainbow serpent, bringer of rain.
Perkunas Virgae	980	-27	162	Lithuanian supreme god, ruler of rain, thunder, and lightning.
Shiwanni Virgae	1400	-25	32	Zuni rain god.
Tishtrya Virgae	276	23.84	179.75	Persian rain god.
Tlaloc Virgae	600	23.73	207.7	Aztec rain god.
Uanui Virgae	917	45.2	235.3	Māori (New Zealand) 'Great Rain' god.

Arc (Arcus), Paterae, Mountains (Montes), Hills (Colles) and Labyrinths				
Hotei Arcus	600	-28	79	Always-smiling god of contentment, good fortune, cheerfulness in Japanese Buddhism. Bright arc of hills bounding 5-micron bright area, once considered possibly volcanic but most likely lakebed deposits.
Arwen Colles	64	-7.5	260	Daughter of Elrond; character from Middle-earth, the fictional setting in fantasy novels by J.R.R. Tolkien.
Bilbo Colles	164	-4.23	38.56	Hobbit, the main character in the eponymous novel by Tolkien; he travelled across Middle-earth.
Faramir Colles	82	4	153.8	Wise man of nobility; character from Middle-earth, the fictional setting in fantasy novels by J.R.R. Tolkien.
Gandalf Colles	102	14.61	209.54	Wizard, leader of the Fellowship of the Ring; character from Middle-earth.
Handir Colles	100	10	356.68	Son of Haldir, a ruler of the fictional Folk of Haleth; character from Tolkien's Middle-earth.
Nimloth Colles	90	11.86	151.3	Second queen of Doriath, a fictional realm of Grey Elves; character from Middle-earth.
Anbus Labyrinthus	120	39.2	215	Planet from the Dune series; IV Anbus, home of Tibana, an apologist for Socratic Christianity.
Corrin Labyrinthus	280	66	21	Planet from the Dune series; site of Battle of Corrin
Ecaz Labyrinthus	000	-0.2	56.7	Planet from the Dune series, useful for its agricultural contributions to the Imperium.

Albedo Features	Dia (km)	Lat (deg)	Lon (deg W)	Name Origin, Feature remarks
Gammu Labyrinthus	115	-77.9	250	Planet from the Dune series; the name for the planet Giedi Prime after the fall of House Harkonnen.
Gamont Labyrinthus	130	56.8	75	Planet from the Dune series; third planet of Niushe, noted for its hedonistic culture.
Gansireed Labyrinthus	300	-69.3	239.3	Planet from the Dune series; with a village named London.
Ginaz Labyrinthus	160	83	261.7	Planet from the Dune series; home of House Ginaz.
Grumman Labyrinthus	540	-35.3	106.8	Planet from the Dune series; where Duncan Idaho first blooded his sword.
Harmonthep Labyrinthus	363	-72.3	101.4	Planet from the Dune series; sixth stop in the Zensunni migration.
Junction Labyrinthus	484	-47.7	215.3	Planet from the Dune series; once controlled by the Spacing Guild, taken over by the Honored Matres as a stronghold for their leader.
Kaitain Labyrinthus	196	52.37	348.66	Planet from the Dune series; the capital of the Imperium. Wide dissected blob in northern midlatitudes, observed on T16.
Kronin Labyrinthus	270	-35.7	96.27	Planet from the Dune series; captured by the Honored Matres.
Lampadas Labyrinthus	445	-81.8	124	Planet from the Dune series; a centre for Bene Gesserit education, also where Miles Teg is trained as a Mentat.
Lankiveil Labyrinthus	450	-48.2	149.5	Planet from the Dune series; where Glossu Rabban, the nephew of Baron Harkonnen, is a Count.
Lernaeus Labyrinthus	167	-83.4	138	Planet from the Dune series; where Miles Teg retires, also the home planet of his Bene Gesserit mother, Lady Janet Roxbrough.
Muritan Labyrinthus	200	-68.8	219.2	Planet from the Dune series; place where the son of a female supplicant to Alia Atreides was killed.
Naraj Labyrinthus	115	-74.2	35.8	Planet from the Dune series; where the son of Farok lost his eyes.
Niushe Labyrinthus	222	75.1	88.1	Planet from the Dune series; planet captured by the Honored Matres.
Palma Labyrinthus	69	-72.4	31	Planet from the Dune series; home to a Bene Gesserit Keep, attacked by the Honored Matres.
Richese Labyrinthus	200	41.8	199	Planet from the Dune series; fourth planet of Eridani A, supreme in machine culture, noted for miniaturisation.
Salusa Labyrinthus	126	45.6	264.2	Planet from the Dune series; Salusa Secundus, used as the Imperial Prison Planet.
Sikun Labyrinthus	175	-77.9	28.9	Planet from the Dune series, home to the useful plant akarso.
Tleilax Labyrinthus	207	-48	16	Planet from the Dune series; home world of the Bene Tleilax.
Tupile Labyrinthus	84	-80.5	32.2	Planet from the Dune series; 'sanctuary planet' for defeated houses of the Imperium.
Angmar Montes	230	-10	221.9	Name of a mountain range from Middle-earth, the fictional setting in fantasy novels by J.R.R. Tolkien.
Dolmed Montes	400	-11.6	216.8	Name of a mountain from Middle-earth.
Doom Mons	63	-14.65	40.42	Name of a mountain from Middle-earth, the fictional setting in fantasy novels by J.R.R. Tolkien. Suspected cryovolcanic feature, connected to Sotra Patera.
Echoriath Montes	930	-7.4	213.8	Name of a mountain range from Middle-earth.
Erebor Mons	50	-4.97	36.23	Name of a mountain from Middle-earth.
Gram Montes	260	-9.9	207.9	Name of a mountain from Middle-earth.
Irensaga Montes	194	-5.68	212.71	Name of a mountain from Middle-earth.
Merlock Montes	200	-8.9	211.8	Name of a mountain from Middle-earth.
Mindolluin Montes	340	-3.3	208.96	Name of a mountain from Middle-earth.

Albedo Features	Dia (km)	Lat (deg)	Lon (deg W)	Name Origin, Feature remarks
Misty Montes	73	56.8	62.44	Name of a mountain range from Middle-earth.
Mithrim Montes	147	-2.16	127.42	Name of a mountain range from Middle-earth. Prominent set of three mountain ridges in Xanadu.
Moria Montes	107	15.05	190.45	Name of three massive peaks, the Mountains of Moria, at the midpoint of the Misty Mountains range from Middle-earth.
Rerir Montes	370	-4.8	212.1	Name of a mountain from Middle-earth.
Taniquetil Montes	130	-3.67	213.26	Name of a mountain from Middle-earth.
Sotra Patera	40	-14.54	40	Norwegian island. Suspected cryovolcanic feature associated with Doom Mons.

Craters, Ring-shaped Features				
Afekan	115	25.8	200.3	New Guinea goddess of creation and knowledge who teaches people how to live correctly.
Beag	27	-34.74	169.55	Celtic/Irish goddess of water, education, and knowledge. Anyone who drinks the water from her well will become wise.
Forseti	145	25.53	10.4	Norse god, the wisest and most eloquent of the Aesir.
Hano	100	40.3	345.1	Bella Coola (northwestern USA and western Canada) goddess of education, knowledge, and magic. She manifested as a shaman so she could teach the people.
Ksa	29	14	65.4	Lakota and Oglala (South Dakota, USA) god of wisdom.
Menrva	392	20.1	87.2	Etruscan goddess of wisdom. Largest crater known on Titan, observed in early ISS images and in T3 radar swath.
Momoy	40	11.6	44.6	Chumash (California, USA) ancestor shaman and goddess of magic, education, knowledge, health and healing.
Mystis	20	0.07	194.86	Greek nymph, a minor deity, nurse of the god Dionysus, who instructed him in the Mysteries.
Selk	80	7	199	Egyptian goddess of knowledge, writing, education, and reptiles. Crater with prominent ejecta deposits in the northwest of Shangri-La. Destination of the Dragonfly mission.
Sinlap	80	11.3	16	Kachin (N. Burma) wise spirit who dwells in the sky and gives wisdom to his worshippers.
Soi	75	24.3	140.9	Melanesian (New Ireland Island, Papua New Guinea) god of wisdom. 'Freshest-looking' and deepest of Titan's craters.
Guabonito	55	-10.9	150.8	Taino Indian (Antilles) sea goddess who taught the use of amulets. Prominent ring in Shangri-La, just west of Xanadu, seen in early Cassini observations.
Nath	95	-30.5	7.7	(ring) Irish goddess of wisdom.
Paxsi	120	5	341.2	(ring) Aymara (Peru, Bolivia) goddess of the moon, education and knowledge.
Veles	45	2	137.3	(ring) Slavic god of housekeeping wisdom.

Flow Features (Fluctus) and Rivers (Flumina)				
Ara Fluctus	70	39.8	118.4	Armenian god famous for his beauty, also known as a resurrected god who dies and is reborn periodically.
Leilah Fluctus	190	50.5	77.8	Persian goddess of beauty and chastity.
Mohini Fluctus	347	-11.78	38.53	Indian goddess of beauty and magic.
Rohe Fluctus	103	47.3	37.75	Beautiful Māori goddess, wife of Māui.
Winia Fluctus	300	49	46	Indonesian first woman, known for her great beauty.
Celadon Flumina	160	-73.7	28.8	River in *The Iliad*.
Elivagar Flumina	260	19.3	78.5	In Norse mythology, 12 poisonous ice-cold rivers. Branching river network east of the Menrva crater.

Albedo Features	Dia (km)	Lat (deg)	Lon (deg W)	Name Origin, Feature remarks
Hubur Flumen	84	-70.24	192.85	The river that flows at the gates of the netherworld in Mesopotamian mythology. Drains into Ontario Lacus.
Karesos Flumen	83	-70.9	194.75	River in *The Iliad*. Drains into Ontario Lacus.
Saraswati Flumen	2.9	-74.55	193.51	A river in Hindu mythology. Drains into Ontario Lacus.
Vid Flumina	158	72.9	242.25	In Norse mythology, a broad river of the ice cold poisonous Elivagar system. Long network of deep canyons draining into Nicoya Sinus and Ligeia Mare, observed with radar imaging and with altimetry.
Xanthus Flumen	78	83.47	242.76	River in *The Iliad*.

Seas (Maria), Islands (Insulae), Inlets (Sinus) and Straits (Freta)				
Kraken Mare	1170	68	310	Fabulous sea monster in the Norwegian seas, said to be a mile and a half in circumference and to cause a whirlpool when it dives. Largest of Titan's seas, comprises two prominent basins connected by Seldon Fretum.
Ligeia Mare	500	79.7	247.9	One of the sirens in Greek mythology. Best-observed of Titan's seas.
Punga Mare	380	85.1	339.7	Māori (New Zealand) supernatural being, the father of sharks and lizards. Son of the sea god Tangaroa.
Bayta Fretum	165	73	311.2	Bayta Darell, fictional character in Isaac Asimov's Foundation series, wife of the trader Tran Darell. Channel at western margin of Kraken Mare, near Penglai and Bimini Insulae; transient reflections observed nearby by Cassini, indicating possible wave- or current-roughening.
Hardin Fretum	246	57.3	317.8	Salvor Hardin, fictional character in Isaac Asimov's Foundation series, first mayor of the planet Terminus. Channel at extreme south of Kraken Mare, linking to Walvis Sinus.
Seldon Fretum	67	66	316.6	Hari Seldon, the fictional, intellectual hero of Isaac Asimov's Foundation series, First Minister of the Galactic Empire. Informally referred to as 'the Throat of Kraken', probable site of major tidal currents.
Trevize Fretum	173	74.4	269.9	Golan Trevize, fictional character in Isaac Asimov's Foundation series, councilman of the planet Terminus. Somewhat grid-like labyrinth of channels joining Ligeia and Kraken Mare.
Bermoothes Insula	124	67.1	317.1	An enchanted island in Shakespeare's *Tempest*.
Bimini Insula	39	73.3	305.4	Boicua; Island in Arawak (Bahamas) legend said to contain the fountain of youth. Small Kraken island, possible wave-roughening nearby.
Bralgu Insulae	55	76.2	251.5	Baralku; in Yolngu culture (Arnhem Land, Australia), the island of the dead and the place where the Djanggawul, the three creator siblings, originated.
Buyan Insula	48	77.3	245.1	A rocky island in Russian folk tales located on the south shore of the Baltic Sea. Well-observed feature with stereo topography. Bounded on north by the shallow and possibly tidal Genova Sinus, and at east by Tunu Sinus.
Hufaidh Insulae	152	67	320.3	Legendary island in the marshes of southern Iraq.
Krocylea Insulae	74	69.1	302.4	Crocvlea; mythological Greek island in the Ionian Sea, near Ithaca. Irregular cluster of islands in eastern Kraken Mare.
Mayda Insula	168	79.1	312.2	Legendary island in the northeastern region of the Atlantic Ocean.
Penglai Insula	94	72.2	308.7	Mythological Chinese mountain Island where immortals lived.

Albedo Features	Dia (km)	Lat (deg)	Lon (deg W)	Name Origin, Feature remarks
Planctae Insulae	64	77.5	251.3	In Greek mythology the 'wandering rocks', between which the sea was mercilessly violent. Only Argo was said to have successfully passed the rocks.
Roylло Insula	103	68.3	297.2	Island in Atlantic Ocean (15th-century maps) on verge of the unknown, near Antilla and St Brandan. Large island on eastern edge of Kraken Mare, between Krocylea Insulae and Gabes Sinus.
Arnar Sinus	101	72.6	322	Fjord in Iceland. Straight-edged bay on western margin of Kraken Mare.
Baffin Sinus	110	80.35	344.62	Bay in Canada. Perhaps better described as a lagoon, an irregular liquid-filled region connected to Kraken via Genova Sinus.
Boni Sinus	54	78.69	345.38	Bone; gulf in Indonesia.
Dingle Sinus	80	81.36	336.44	Bay in Ireland.
Flensborg Sinus	115	64.9	295.3	Fjord between Denmark and Germany. Southeastern corner of the main (northern) basin of Kraken Mare.
Gabes Sinus	147	67.6	289.6	Gulf in Tunisia.
Genova Sinus	125	80.11	326.61	Gulf in Italy. Possible tidal flat behind Mayda insula. Connects (via Baffin Sinus) Punga Mare to Kraken.
Kumbaru Sinus	122	56.8	303.8	Bay in India. Southeastern lobe of Kraken Mare.
Maizuru Sinus	92	78.9	352.53	Bay in Japan.
Manza Sinus	37	79.29	346.1	Bay in Tanzania.
Moray Sinus	204	76.6	281.4	Firth in Scotland. Inlet at northeast corner of Kraken Mare – apparent surface reflectivity changes have been observed suggesting waves or strong currents.
Nicoya Sinus	130	74.8	251.2	Gulf in Costa Rica. Southern margin of Ligeia Mare, where Vid Flumina enters.
Okahu Sinus	141	73.7	282	Bay in New Zealand. Inlet in northeastern corner of Kraken, mouth of Trevize Fretum.
Patos Sinus	103	77.2	224.8	Fjord in Chile. Northeastern corner of Ligeia Mare.
Puget Sinus	93	82.4	241.1	Sound in state of Washington, USA. Notable inlet in north margin of Ligeia Mare, features meandering river and possible relict impact crater.
Rombaken Sinus	92.5	75.3	232.9	Fjord in Norway. Deeply incised bay on south margin of eastern lobe of Ligeia Mare.
Skelton Sinus	73	76.8	314.9	Inlet in Antarctica.
Trold Sinus	118	71.3	292.7	Fjord in Canada. Heavily incised bay at eastern margin of Kraken Mare.
Tunu Sinus	134	79.2	299.8	Fjord in Greenland.
Wakasa Sinus	146	80.7	270	Bay in Japan. Western lobe of Ligeia Mare, with much less irregular shoreline than Rombaken Sinus.
Walvis Sinus	253	58.2	324.1	Bay in Namibia. Irregular lobe at extreme southern margin of Kraken Mare, to which it connects via Hardin Fretum.

Lakes (Lacus) and possible Lakebeds (Lacunae)				
Atacama Lacuna	35.9	68.2	227.6	Intermittent lake (salar) in Chile.
Eyre Lacuna	25.4	72.6	225.1	Intermittent lake in Australia.
Jerid Lacuna	42.6	66.7	221	Intermittent lake (shott) in Tunisia.
Kutch Lacuna	175	88.4	217	Intermittent lake on the border of India and Pakistan.
Melrhir Lacuna	23	64.9	212.6	Intermittent lake (shott) in Algeria.
Nakuru Lacuna	188	65.81	94	Intermittent lake in Kenya.

Albedo Features	Dia (km)	Lat (deg)	Lon (deg W)	Name Origin, Feature remarks
Ngami Lacuna	37.2	66.7	213.9	Intermittent lake in Botswana.
Racetrack Lacuna	9.9	66.1	224.9	Intermittent lake (playa) in California, USA.
Uyuni Lacuna	27	66.3	228.4	Intermittent lake (salar) in Bolivia.
Veliko Lacuna	90	-76.8	33.1	Intermittent lake in Bosnia-Herzegovina.
Woytchugga Lacuna	449	68.88	109	Intermittent lake in Australia.
Abaya Lacus	65	73.17	45.55	Lake in Ethiopia. 'Kissing Lakes' – two prominent lobes joined at a narrow point.
Akmena Lacus	35.6	85.1	55.6	Lake in Lithuania.
Albano Lacus	6.2	65.9	236.4	Lake in Italy.
Annecy Lacus	20	76.8	128.9	Lake in France.
Arala Lacus	12.3	78.1	124.9	Lake in Mali.
Atitlan Lacus	13.7	69.3	238.8	Lake in Guatemala.
Balaton Lacus	35.6	82.9	87.5	Lake in Hungary.
Bolsena Lacus	101	75.75	10.28	Lake in Italy. Large elliptical lake.
Brienz Lacus	50.6	85.3	43.8	Lake in Switzerland.
Buada Lacus	10.3	76.4	129.6	Lake in Nauru.
Cardiel Lacus	22	70.2	206.5	Lake in Argentina.
Cayuga Lacus	22.7	69.8	230	Lake in New York, USA.
Chilwa Lacus	19.8	75	131.3	Lake in Malawi and Mozambique.
Crveno Lacus	41	-79.55	184.91	Lake in Croatia. Second-best observed of the few southern lakes. Elliptical lake 350km south of Ontario Lacus.
Dilolo Lacus	18.3	76.2	125	Lake in Angola.
Dridzis Lacus	50	78.9	131.3	Lake in Latvia.
Feia Lacus	47	73.7	64.41	Lake in Brazil.
Fogo Lacus	32.3	81.9	98	Lake in Portugal, Azores.
Freeman Lacus	26	73.6	211.1	Lake in Indiana, USA.
Grasmere Lacus	33.3	72.3	103.1	Lake in England.
Hammar Lacus	200	48.6	308.29	Lake in Iraq. Somewhat isolated lake (400km south of Kraken Mare) – southernmost northern lake?
Hlawga Lacus	40.3	76.6	103.6	Lake in Myanmar.
Ihotry Lacus	37.5	76.1	137.2	Lake in Madagascar.
Imogene Lacus	38	71.1	111.8	Lake in Idaho, USA.
Jingpo Lacus	240	73	336	Lake in China (means 'Mirror Lake'). So named for the prominent specular reflection of the sun observed from this feature by VIMS on T58.
Junin Lacus	6.3	66.9	236.9	Lake in Peru.
Karakul Lacus	18.4	86.3	56.6	Lake in Tajikistan.
Kayangan Lacus	6.2	-86.3	202.17	Lake in the Philippines. Small lake near south pole.
Kivu Lacus	77.5	87	121	Lake on the border between Rwanda and the Democratic Republic of the Congo. Lake closest to Titan's north pole.
Koitere Lacus	68	79.4	36.14	Lake in Finland. Irregular-shaped lake with island, near Neagh Lacus.
Ladoga Lacus	110	74.8	26.1	Lake in Russia.
Lagdo Lacus	37.8	75.5	125.7	Lake in Cameroon.
Lanao Lacus	34.5	71	217.7	Lake in the Philippines.
Letas Lacus	23.7	81.3	88.2	Lake in Vanuatu.
Logtak Lacus	14.3	70.8	226.1	Lake in Manipur, India.
Mackay Lacus	180	78.32	97.53	Lake in Australia. Very irregular-shaped lake with multiple islands.
Maracaibo Lacus	20.4	75.3	127.7	Lake in Venezuela.

Albedo Features	Dia (km)	Lat (deg)	Lon (deg W)	Name Origin, Feature remarks
Muggel Lacus	170	84.44	203.5	Lake in Germany. Name suggested by a Harry Potter fan.
Muzhwi Lacus	36	74.8	126.3	Muzhwi Dam, lake in Zimbabwe.
Mweru Lacus	20.6	71.9	131.8	Lake in Zambia and Democratic Republic of the Congo.
Myvatn Lacus	55	78.19	135.28	Lake in Iceland.
Neagh Lacus	98	81.11	32.16	Lake in Northern Ireland, United Kingdom. Irregular lake with large island(s).
Negra Lacus	15.3	75.5	128.9	Lake in Uruguay.
Ohrid Lacus	17.3	71.8	221.9	Lake on the border of Macedonia and Albania.
Olomega Lacus	15.7	78.7	122.2	Lake in El Salvador.
Oneida Lacus	51	76.14	131.83	Lake in New York, USA. Crescent-shaped lake, observed to be approx 100m deep.
Ontario Lacus	235	-72	183	Lake on the border between Canada and the United States. Largest lake in the southern hemisphere, suspected of being ethane-rich dried-up remnant of a once-larger lake.
Phewa Lacus	12	72.2	124	Lake in Nepal.
Prespa Lacus	43.7	73.1	135.7	Lake in the Republic of Macedonia, Albania, and Greece.
Qinghai Lacus	44.3	83.4	51.5	Kukunor, Tso Ngonpo; lake in China.
Quilotoa Lacus	11.8	80.3	120.1	Lake in Ecuador.
Rannoch Lacus	63.5	74.2	129.3	Lake in Scotland.
Roca Lacus	46	79.8	123.5	Lake in Chile and Argentina.
Rukwa Lacus	36	74.8	134.8	Lake in Tanzania.
Rwegura Lacus	21.7	71.5	105.2	Lake in Burundi.
Sevan Lacus	46.9	69.7	225.6	Lake in Armenia.
Shoji Lacus	5.8	-79.74	166.37	Lake in Japan.
Sionascaig Lacus	143.2	-41.52	278.12	Lake in Scotland.
Sotonera Lacus	63	76.75	17.49	Lake in Spain.
Sparrow Lacus	81.4	84.3	64.7	Lake in Canada. Rather elongate in shape.
Suwa Lacus	12	74.1	135.2	Lake in Japan.
Synevyr Lacus	36	81	53.6	Lake in Ukraine.
Taupo Lacus	27	72.7	132.6	Lake in New Zealand.
Tengiz Lacus	70	73.2	105.6	Lake in Kazakhstan.
Toba Lacus	23.6	70.9	108.1	Lake in Indonesia.
Towada Lacus	24	71.4	244.2	Lake in Japan.
Trichonida Lacus	31.5	81.3	65.3	Lake in Greece.
Tsomgo Lacus	59	-86.37	162.41	Tsongmo, Changu; lake in India. Southernmost of the southern hemisphere lakes.
Urmia Lacus	28.6	-39.27	276.55	Lake in Iran.
Uvs Lacus	26.9	69.6	245.7	Lake in Mongolia. Elliptical lake south of Ligeia Mare.
Vanern Lacus	43.9	70.4	223.1	Lake in Sweden. Irregular lake south of Ligeia Mare.
Van Lacus	32.7	74.2	137.3	Lake in Turkey.
Viedma Lacus	42	72	125.7	Lake in Argentina.
Waikare Lacus	52.5	81.6	126	Lake in New Zealand.
Winnipeg Lacus	60	78.05	153.31	Lake in Canada. Seahorse-shaped lake, measured by radar on T126 to be methane-rich and ~100m deep.
Xolotlan Lacus	57.4	82.3	72.9	Managua; lake in Nicaragua.
Yessey Lacus	24.5	73	110.8	Lake in Siberia (Evenkia, Asiatic Russia).
Yojoa Lacus	58.3	78.1	54.1	Lake in Honduras.
Ypoa Lacus	39.2	73.4	132.2	Lake in Paraguay.
Zaza Lacus	29	72.4	106.9	Lake in Cuba.
Zub Lacus	19.5	71.7	102.6	Lake in Antarctica.

Further reading

My own account of the state of our knowledge about Titan prior to the Cassini mission is *Lifting Titan's Veil*, R. Lorenz and J. Mitton, Cambridge University Press, 2002. The story was updated, with first-hand anecdotes about Cassini's first several years of discoveries, in *Titan Unveiled*, R. Lorenz and J. Mitton, Princeton University Press, 2010.

Titan: Exploring an Earth-Like World, A. Coustenis and F. Taylor, World Scientific, 2008, is a slightly higher-level book. But the most comprehensive academic summary of the findings at Titan of Cassini's nominal mission (i.e. the first four years) written by dozens of scientists involved, is *Titan from Cassini-Huygens*, R. Brown, et al., Springer 2009. These books do not cover the last decade of the Cassini mission, however. A more up to date compilation is *Titan: Interior, Surface, Atmosphere, and Space Environment*, I. Mueller-Wodarg et al., Cambridge University Press, 2014.

Beyond Earth, A. Hendrix and C. Wolforth, Pantheon Books, 2016, takes a look at whether (and how) Titan may be a destination for future human colonisation.

Among science fiction focused on Titan, *On the Shores of Titan's Farthest Sea*, M. Carroll, Springer, 2015, is the most fun so far.

The Planetary Photojournal (http://photojournal.jpl.nasa.gov/) has a good archive of the best images from Cassini and other missions, but for the full selection of raw data and images the place to visit is the Planetary Data System (PDS):
https://atmos.nmsu.edu/data_and_services/atmospheres_data/Cassini/sci-titan.html

Details on the Dragonfly mission can be found at:
http://dragonfly.jhuapl.edu

BELOW Cassini's last radar imaging of Titan in April 2017 shows the coast of Ligeia Mare, the Vid Flumina canyon system, and the lakes (Laci) Atitlán, Uvs and Towalda. *(Author)*

Index

Airglow 6, 138, 141
Amino acids 170, 173
APL 153, 155, 158-160, 162
Appearance 23-24
  colour 23
  disc 13, 23-24, 27, 33
  lightcurve 28
  Mercator views 42
  nightside 25
  winter darkness 100
Asimov, Isaac 59
Atmosphere – throughout
  chemistry 15, 50, 63, 135, 137-151, 153, 167, 174, 178
  composition – see also Gases 120-122, 149
  condensate ices 26
  density 25, 38-40, 68, 82, 128, 174, 177
  dust storms 127, 130
  electrons 143
  energetic neutral atoms (ENA) 139, 142
  haze 13, 25, 27-28, 31, 41, 47-48, 50, 124, 131, 134-135, 144-145, 148
  ionosphere 138, 141-143
  layers 140, 149
  light levels 41, 48
    year-to-year variations 29
  lower and surface atmosphere 143, 150
  meridional circulation 124
  molecules and ions 145, 148
  oxygen atoms 151
  planetary boundary layer (PBL) 96
  pressure 176
  radiocarbon 144
  scattering of light 130, 140
  solar ultraviolet radiation 144
  temperature 24, 161-62
  thickness 25, 40, 71, 114, 130, 144, 154
  upper 45, 48, 67, 131, 139, 142, 144, 167, 170
  visibility 47

Balloons and airships (lighter-than-air travel) 154-157, 175-176
Baxter, Stephen 124
Boat concepts 160, 177

Calendar of observations and events 19
Carbon and hydrological inventories 145
Canyonlands 82, 86-87, 90
Cassini, Giovanni Domenico (Jean-Dominique) 22
Cassini-Huygens mission – see also Huygens
  probe 6, 13, 19, 21, 26-27, 31, 126
  duration 151
  Equinox Mission 50
  extended 49-50
  fuel supplies 49-50
  insertion to orbit around Saturn 39-40
  launch 37, 154
  thruster propellant 50
  Titan flybys 39-40, 48-50, 64-66, 71, 75, 143, 151, 156
  trajectory 37
  voyage 35, 37, 49

Cassini-Huygens systems
  Cassini Plasma Spectrometer (CAPS) 39, 49, 147-148
  Composite InfraRed Spectrometer (CIRS) 39, 46, 133, 149, 166
  Cosmic Dust Analyser (CDA) 39
  Descent Imager/Spectral Radiometer (DISR) 41-42
  design life 50
  Gas Chromatograph/Mass Spectrometer (GCMS) 43, 62
  Imaging Science Subsystem (ISS) 30, 39
  instruments 37, 39-41, 49, 100, 138, 149
  Ion and Neutral Mass Spectrometer (INMS) 39, 62, 147-148, 158
  Magnetospheric Imaging Instrument (MIMI) 39, 142
  Optical Remote Sensing (ORS) 39
  radar 38, 75, 82, 128, 181-183
  Radio and Plasma Wave System (RPWS) 39
  Radioisotope Thermoelectric Generators (RTG) 49
  Radio Science Subsystem 50
  radio equipment 39
  Ultraviolet Imaging Spectrometer (UVIS) 38, 148
  Visual and Infrared Mapping Spectrometer (VIMS) 38-39, 166
Clarke, Arthur C. 17, 56, 126, 174
Climate 9-11, 104, 165, 178
Clouds (methane) 6, 10, 19, 28, 31, 33, 40, 42, 94, 100, 119-120, 122-128, 131, 133
  ethane 132
  polar vortex 134
Cosmic rays 142-144
Cryovolcanism 70, 170, 178
Curiosity rover 160-162, 165-166

Data 18, 48, 166-167
  diameter 79
  mass 79
  radius 65
Data transmission directly to Earth 158-162, 164
Days (Tsoi) 158-159, 164
Discovery of Titan 21-22
DNA 165, 171, 173
Dragonfly octocopter-lander mission 6, 16, 18, 73, 85, 92, 97, 149, 153, 160-167
  arrival 166
  exploration strategy 164
  flying range 164
  instruments, sensors and tools 163-166
  Iron Bird frame model 163
  landing sites 164
  launch 166
  rotors 162
  sampling system 163-165
Dunes and dunefields 5-6, 13, 15, 17-18, 35, 49-52, 57, 70, 73, 76, 78, 81, 91-97, 153, 164-165
  morphology 6
  sands 91, 97, 127, 147, 165

sand seas 68, 92, 100
Shangri-La 6, 53, 78

Earth 9-11, 26, 71
  atmosphere 10, 25, 62, 120, 138, 166
  aurorae 138, 141
  carbon dioxide 170
  climate 11, 104
  clouds 131-132
  craters 71-74
  Death Valley, California 83-84, 102
  deserts 82, 91, 93, 97
  drones and UAVs 156, 160
  equator 128
  flyby 37
  geysers 57
  gravity 175
  hot air ballooning 155
  hydrogen 170
  hydrological cycle 10
  ice ages 104
  impact craters 141, 171-172
  jetstream 45, 126
  life 16, 178
  lithosphere (crust) 66
  magnetic field 138
  matter from space 150, 171
  mesosphere 150
  methane 170
  Neumayer station, Antarctica 177
  nickel deposits 171
  oceans, seas and lakes 46, 79, 101-103, 109, 111, 117
  optical image 32
  oxygen 170
  ozone 23; hole 45, 133
  rotation(spinning) 94, 111-112, 124
  salt glaciers 55, 57, 86
  sand dunes and seas 81, 92, 94-97
  sands 91, 97
  seasons 104
  stratosphere 131-132, 150
  surface 77
  radar reflectivity 59
  tidal flows 113
  topography 77
  transition to living matter 171-172
  warfare 156
  water 11, 124, 178
  vapour 46, 121, 170
  weather 121
Equator 11, 67, 70, 92, 100, 126, 128-129
  tilt 10-11
Equinox 31, 96, 121, 124-127, 130, 13
  northern spring 28, 119
  spring 132
Erosion 66, 68, 82, 90, 105, 116
European Space Agency (ESA) 26, 157
Exomoons 11
Exoplanets 9, 11

Formation of Titan 79

Galileo 62
Galileo mission 37, 64, 150
Garry, James 155
Gases and organic compounds 23, 25-26, 56, 133, 170
  acetylene 25, 43, 47
  acrylonitrile 167
  ammonia 62-63
  argon 40, 62
  carbon dioxide 25, 43, 150
  carbon monoxide 33, 138, 176
  cyanogen 25
  ethane 43, 106, 117, 147
  hydrogen 24, 46-47, 121, 138, 145
  hydrogen cyanide (HCN) 132, 134, 167, 176
  hydrogen isocyanide (HNC) 167
  methane 11, 23-24, 26-28, 40, 43, 46-47, 50, 106, 116-117, 121, 126, 128, 134, 138, 141, 144-145, 147, 161, 178
  molecular nitrogen 10, 25
  nitrogen 40, 47, 62-63, 89, 116-117, 121, 128, 142, 144, 160-161
  nitriles 33, 45, 47, 132, 147, 167
  propane 25, 89
Geological particle sizes 91
Gravity 18, 39, 41, 44, 47, 49, 65-66, 82, 86, 93, 111, 114-115, 124, 151, 154, 174-176
Greenhouse effect and gases 120-122, 145

Halley's comet 26
Heavier-than-air transport – see also Dragonfly 154-157, 176
  AVIATR concept 158
  flight control and navigation 162
  helicopters 154, 160
  radioisotope thermoelectric generators (RTGs) 155
  Titan Bumblebee 16
Hendrix, Amanda 177
Herbert, Frank 59
Hotei Regio lakebed 53-56, 78
Hubble Space Telescope (HST) 28-33, 77, 138, 166
Humans (astronauts) on Titan 174-178
  bipedal walking 175-176
  habitats 176-177
  insulated suits 176
  oxygen supply 176
Huygens, Christiaan 22, 79
Huygens probe 6, 15-16, 40 *et seq.*
  entry and descent 31-33, 35, 37, 40-48, 68, 82, 127, 130
  landing site 42-43, 48, 51, 56, 71, 77, 82, 100, 154
  radio equipment 40
  spectrometers 15-16
  telecommunications system 32
  touchdown (impact) 35, 37, 43
Hydrological cycles 10-11, 13, 30

Ice crust 61, 64, 66-67, 77, 82, 112, 147, 166, 172, 178
  Titanquakes 166
Images and data 15, 28-29
  descent 43
  DISR 47
  hand-held camera 93
  Huygens probe 15, 35

HST 38, 138
infrared 135
ISS 30, 38, 41-42, 55, 75, 100, 108
mosaiced 37, 40, 48, 55, 58, 72, 99-100, 103, 108, 182-183
near-infrared 5-6, 13, 15, 27, 100, 121-122, 134, 167
optical 82
Pioneer 11 24
radar 15, 30, 38, 50, 53, 55, 58, 68-69, 71, 77, 79, 82, 84, 87, 90, 92-93, 95, 97, 99, 103, 107, 110, 115, 181-183
SAR 78
Space Shuttle 84, 93, 103
Space Station 95
true (natural) colour 9, 15, 135
ultraviolet 134-135, 138, 149
Venus Express 135
VIMS 38, 41, 43, 51, 53-55, 58, 68-69, 71, 73, 85, 92, 102, 104, 115, 127, 130, 132, 141, 171, 181
Impact craters 15, 18, 49, 51-53, 57, 61, 70-78, 119, 153, 171, 188
  Menrva 82
  Selk 56, 73, 97, 167, 171
Interior of Titan 13, 61, 65, 147, 169, 172
International Astronomical Union 15
  Committee on Planetary Nomenclature 59
International Space Station 95, 141

James Webb Space Telescope (JWST) 166-167
JET (Journey to Enceladus and Titan) mission concept 158
Jovian (Jupiter's) moons (Galileans) 62, 64, 79
  Callisto 62-63
  Europa 16, 62-63, 77, 82, 158, 172
  Ganymede 9-10, 62-63, 75-76, 150, 178
  Io 62
JPL (Jet Propulsion Laboratory) 154, 157-158, 160
Jupiter 10-11, 24, 37, 62, 64, 151, 160
  flyby 157
  satellites 10

Kuiper, Gerard 23

Lakes and seas (methane) 5, 11, 15, 39, 46, 50, 52, 56-57, 59, 79, 91, 96, 99-117, 122, 124, 151, 169, 173, 177, 182-183, 189-190
  bathtub rings 100-102
  depths 107-108, 112-113, 157
  evaporation 120, 158
  internal water ocean 63, 66, 169, 172, 178
  Jingpo Lacus 15, 114
  Kraken Mare 16-17, 104, 107-108, 110, 112-113, 115-116, 157, 159
  Ligeia Mare 16, 68, 87, 91, 99, 104-105, 107-108, 112, 115-117, 158, 175
  Ontario Lacus 50, 56, 100, 102-105, 112
  open seas 100, 144
  Punga Mare 104, 109-110, 112, 115-116
  radio transparency 159
  sea levels 99, 109, 113
  tides 111 113, 151, 156
  waves 114-117, 157-159
Landscape 9, 81-82, 119, 153
Latitude 56, 64, 77, 122, 124, 126, 128-129, 166

Life on Titan? 169-178
  building blocks 170, 174
  chemistry 173-174
  weird 47
Light and radio signals from Earth 16, 162
Liquids 25-26, 178
  ethane 50, 67, 89, 100, 106, 124, 173
  hydrocarbon 182-183
  methane 25, 82, 89, 99-100, 106, 108, 120-121, 124, 160, 169, 172, 174, 177-178
  surface tension 124
  physical properties 106-107
  water 61, 169, 172
Lockheed Martin 158
Longitude 56, 64
Lorenz, Ralph (author) 101, 114, 124, 154-155, 166

Maps and mapping 15, 26, 28, 30, 35, 37-38, 48, 50-51, 54-56, 156, 180-183
  ALMA 167
  Cassini ISS 53, 96, 125, 171
  dune orientation 96
  Hubble 77
  HST 91
  near-infrared 30, 130
  RADAR 30, 48, 58
  radiometry 52
  topographic 67
  winds 167
Mars 9-11, 16, 37, 43, 62, 68, 77, 172
  atmosphere 24, 62, 120
  Beagle 2 lander 155, 166
  exploration and missions 154, 157-158
  Phoenix lander 155
  sands 91
  seasons 104
  sky 130
Materials strengths (rocks and ice) 88-89
Mercury 9, 62
Meteors, meteorites and meteoroids 73, 138, 142, 140, 150, 170
Microwave emissivity 51-52, 75, 79
Moon 30, 51, 67-68, 71, 76, 79, 82, 111-112, 139
  Apollo lunar rover 176
  Apollo moonwalkers 175-176
  gravity 175-176
  habitats 176
Mountains and hills 30, 53, 55, 66, 68-70, 78, 82, 85, 186
Musgrave, Astronaut Story 93

Names of Titan's features and places 15, 59, 184-
  Albedo features 59, 184-192
  regions 185
NASA 6, 16, 26, 154-155, 157, 160, 166
  Ames Mars Wind Tunnel 114
  CAESAR comet sample mission 160
  Decadal Survey committee 158
  Discovery program 158
  Innovative Advanced Concepts program 159
  New Frontiers program 6, 153, 160, 166
  New Horizon mission to Pluto 145, 160
    Titan flyby 145
  OSIRIS-Rex asteroid mission 160

Neptune 11, 151
  Triton moon 79, 88
Northern hemisphere 24-25, 29, 50, 103-104, 122, 132, 178

OCEANUS Titan mission concept 157, 160
Orbit around Saturn 10-11, 22, 39, 49, 64-65, 79, 111-113, 119, 130, 138, 151, 154, 178

Paris Observatory 22
Parker Solar Probe 161
Photochemistry 67-68, 120, 131, 144-150, 170
Pioneer 11 24
Plains 52, 57, 70, 185
  cobbles 82, 84-85
Planetary science 23
Plasma 138
Pluto 9, 88, 145, 150
  atmosphere 150
  chemistry 150
Pocket watch lander 155
Polar regions 49, 64-65, 67, 96, 100, 122, 130, 178
  hood 133, 135
  north pole 5, 11, 15, 21, 25, 55, 97, 132, 134, 149, 183
  south pole 6, 29-30, 40, 90, 124, 126, 132-134, 182
Polycyclic aromatic hydrocarbons (PAH) 146-147, 169

Radio isotope power sources 155-157, 161
  ASRG power units 157-158
  MMRTG (MultiMission Radioisotope Thermoelectric Generators) 155-156, 160-161, 163
  radio isotope generators (RTGs) 155
Rainfall (methane) 10-11, 30, 42, 75, 77, 82-83, 96, 104, 109, 119, 122-128
  Bond Number 124
  condensation 123
  raindrop speed 127
Rivers, streams and channels 11, 15, 42, 51, 53-54, 68, 75, 78, 81-90, 95, 97, 105, 176, 188
  methane 82
  sediment 82, 85
Robotic spacecraft 23
Rotation (spinning) 10, 28, 64-65, 67, 94, 96, 111-112, 124, 130

Sagan, Carl 23-25
Saturn – throughout
  appearance form Earth 22
  aurorae 138
  equator 64-65
  gravity 39, 63, 112, 138, 151
  magnetic field 64, 138-139, 143, 151, 170
  magnetosphere 25, 138-140, 142-143
  orbit around the Sun 13, 22, 79, 95, 140
  rings 9, 13, 22, 26, 31, 40, 50, 130, 137, 150-151
    Colombo Gap 151
  Saturnshine 131
  seasons 22

Saturnian system (moons) 10-11, 22, 24, 31, 33, 49-50, 58, 79, 97, 137, 140, 150-151
  Dione 75
  Enceladus 39, 63, 71, 137-138, 150-151, 155, 157-158, 170
  Hyperion 151
  Iapetus 75
  Janus 132
  Phoebe 40, 7xxx, 79, 151
  Rhea 79
Science fiction novels 16-17
Seasons 10-11, 13, 22, 28-29, 39, 55, 104, 110, 119, 122, 135, 138
  northern summer 132, 158
  polar summer 128
  polar winter 128, 131
  southern summer 132, 178
  spring 28, 119, 122
  summer 115, 124, 129
  summer hemisphere 131
  winter hemisphere 13
Sky 130
Solà, Josep Comas 22-24
Sola electric propulsion 157
Solar system 9, 62, 79, 99, 172
  age 26
  climates 178
  moons 10
Southern hemisphere 25, 29, 50, 103, 117, 134
Spacecraft descents and landings – see also Huygens probe 16, 18, 27, 37, 155-157
  hypersonic entry into atmosphere 161
  parachutes 16, 27, 37, 40-41, 43, 45, 48, 68, 157, 161, 164
  splashdowns 157-159, 174
Stars
  starlight 38
  TYC 1343-1615-1 27
  28 Sagittarii 26
Stirling generators 158-159
Stratosphere 19, 23, 26, 31, 120, 131
  circulation 131-135
  temperatures 44, 46
Submarine concept 159
Sun 11, 13, 22, 25, 39, 46, 62, 96, 104, 110-112, 130-131, 140, 170, 178
  charged particles 137-138
  evolution to red giant 145, 178
  reflection on sea surface 114-115
  solar aureole 47
  solar winds 138-139, 151
  sunlight 37-38, 115, 121-122, 131, 140-141, 161, 178
  Titan's travels around the sun 130
    eclipse 131, 140
  ultraviolet light 46, 137-138, 141-142, 150, 170
  X-rays 138
Supernovae 138
Surface 13, 15, 26, 28-33, 37, 39-43, 50, 52, 61, 71-72, 76-77, 100, 120, 128-130, 154, 156, 172
  chemistry 154, 165
  material analysis 154
  pressure 26
  radar reflectivity 30, 39
  subsurface 41

temperatures 24-26, 41, 84, 100, 124, 128-130, 178
trafficability 154, 176

Tectronics 68-70
Telescopes – see also Space Telescopes HST and JWST 22-23, 27-28, 30, 33, 45, 64, 124, 167
  ALMA 167
  Arecibo 30, 32
  Cassini 41
  Goldstone 26, 30
  Green Bank 44
  ground-based 30, 38, 44, 166
  IRAM 33
  Keck II 31, 53
  Lowell Observatory 135, 166
  near-infrared 131
  Parkes 44
Terrain 33, 57-58, 66, 90
Terrestrial planets 9-10
  climates 10
Thermal infra-red heat radiation 23
Tholins 146, 148-149
Titan Explorer Flagship Mission Study 155-158
Titan Mare Explorer Mission (TIME) 129, 158-159
Titan Saturn System Mission (TSSM) 157-159
  lake lander 157
  orbiter 157-158
  Titan flyby 157
Tolkien, J.R.R. 59, 69
Topography 17, 30, 41, 44, 54, 66-68, 70
Troposphere 19, 26, 28, 31, 120
  temperatures 44
TV documentaries 77

United Nations Treaty 6
Uranus 10-11, 79, 151
  moons 10, 79

Venus 9-11, 16, 26, 43, 46, 63, 71, 77, 120, 134
  Akatsuki mission 162
  equator 11
  gravity 37
  orbital plane 11
  rotation 10
Vonnegut, Kurt 101
Voyager 1 and 2 missions 21, 24-30, 33, 40, 45, 68, 76, 120, 128, 131-132, 138, 147

Water ice 51, 62, 67, 73, 75, 88, 172
Weather 28, 67
Winds 26-27, 33, 44, 48, 91, 95-97, 110, 114-116, 119, 127-130, 158, 165-167, 177-178
  circumpolar jet 133
  computer models 159
  stratospheric 45, 131, 133
Wohiforth, Charles 177
Wohler, Friedrich 148

Xanadu 6, 15, 30, 32, 50-53, 56, 59, 68, 71, 77-79, 83, 85, 97

Zubrin, Robert 174